Natural Fragrances

Natural
Fragrances

Outdoor Scents for
Indoor Uses

GAIL DUFF

A Garden Way Publishing Book

STOREY

Storey Communications, Inc.
Pownal, Vermont 05261

First published in the US by Storey Communications, Inc.
Schoolhouse Road
Pownal, Vermont
05261

Library of Congress Cataloging-in-Publication Data

Duff, Gail.
Natural fragrances: outdoor scents for indoor
uses/Gail Duff.
p.   cm,
ISBN 0-88266-554-5: $22.95
1. Pot-pourris (Scented floral mixtures). 2. Handicraft.
I. Title. TT899.4.D84 1989
745.92 – dc19   89-45219
CIP

Conceived and produced by Breslich & Foss
Golden House, 28-31 Great Pulteney Street
London W1R 3DD

Designed by Peartree Design Associates
Typeset by Angel Graphics
Printed and bound in Spain by Graficas Reunidas

CONTENTS

INTRODUCTION

The sweet fragrance of flowers, the fresh, clean scents of herbs and the rich, heady aromas of spices, woods and barks are timeless. Man has for thousands of years, regarded these natural fragrances as being clean and pure, while automatically disliking musty, damp or decaying smells.

In ancient Sudan in 3000 BC frankincense, burned in places of worship, was second only to gold as the most valued article of commerce. In ancient Egypt, at the same period, fragrant preparations of flowers and spices – the first pot-pourris – were placed in tombs and were sufficiently long-lasting as to scent the air heavily when the tombs were finally excavated in the early twentieth century.

Gardens of sweet-smelling flowers and herbs have always been sources of pleasure, relaxation and healing. In ancient Egypt they were situated around pools containing lotus plants, in medieval Europe gardens were enclosed by walls and in the sixteenth century formal knot-gardens were the fashion.

The art of drying flowers and herbs or preparing them in other ways to ensure a lasting scent for the home has probably been practised for as long as they have been cultivated and, although it was almost forgotten in the middle part of the twentieth century, it has regained popularity and is now flourishing.

These days, gardens are blooming with flowers and herbs and we are finding time to enjoy their scents and to discover the old methods of bringing them into our homes. For a few hours each week our kitchen tables can become centers of fragrant industry while we produce pot-pourris and scented candles and pomanders. Keeping these natural fragrances around the house will enable us to experience their timeless joy all year long.

HERBS AND FLOWERS

Herbs and flowers have been cherished for their fragrance for many thousands of years. While they are growing their scents will softly pervade the garden, and, once the plants are dried or distilled into oils, they can be preserved for years.

Nowadays we tend to pigeon-hole plants both in our minds and in our gardens, making a rule that herbs are grown in the herb garden and flowers in the flowerbed. However, you can create a delightful garden by combining them in a border or bed according to their appearance, height and fragrance.

The scent from all fragrant plants is produced by an essential oil, found either in the flowers themselves or, as in the case of herbs, in the leaves. The essential oil from flowers is known as an attar and is made up of a number of chemical compounds, the formula for each particular flower being unique. The essential oils of herbs are simpler, often being made up of only one basic ingredient. Therefore, grown together they can complement each other, the flowers providing stronger, more subtle scents for a shorter time and the herbs giving aromatic undertones that last throughout the summer months.

Herbs have many uses, medicinal, culinary and aromatic, besides making an attractive display of foliage. Flowers have been grown mainly for pleasure, for their appearance and their scents, yet their role too could be called medicinal, for walking or working in a fragrant garden is highly therapeutic.

Indeed, working among sweetly-scented plants is never a chore. Whatever the job you are doing, be it weeding the herbs or removing full-blown heads from the roses, there will always be fragrances around you that you will remember for years to come, each a pleasant reminder of the time of year and the enjoyable tasks that are to be done.

AROMATIC HERBS

The majority of herbs, whether grown for their scent, for the kitchen or for medicinal uses, are highly aromatic and can fill the garden with fragrance besides making a wonderful show of foliage and flowers. Listed below are a selection of herbs to consider.

ALECOST *(Chrysanthemum balsamita)*, also called Costmary: A perennial herb with long, spear-shaped, grey-green leaves and tiny yellow flowers. It has a bitter, spearmint-like scent and can be used in pot-pourris and sweet-bag mixtures.

ANGELICA *(Angelica archangelica)*: An annual plant which can be treated as a biennial or perennial as it readily seeds itself. The thick stems and large, lush, green leaves have a pungent, sweet scent. They can be used in pot-pourris and are often ingredients in toilet waters.

BASIL *(Ocimum basilicum)*: An annual plant best sown indoors in the spring and planted out late in the season. It has a sharp, spicy scent and is occasionally used in pot-pourris.

BAY *(Laurus nobilis)*: Evergreen bay trees can be grown in tubs or in a sheltered spot in the garden. The shiny leaves make excellent additions to pot-pourris and sweet-bag mixtures.

BERGAMOT OR BEE BALM *(Monarda didyma)*: A perennial plant with a sharp scent and red or pink flowers that are a delight to behold. Both leaves and flowers can be added to pot-pourris.

CATMINT *(Nepeta cataria)*: A strongly scented plant that grows nearly 2 ft (60 cm) high and has heads of bright blue flowers. It is pleasant in the herb garden on a sunny day and cats can never resist it.

CHAMOMILE *(Anthemis nobilis)*: A perennial, low-growing, fruity-scented, daisy-like plant with filigree leaves. The flowers are added to pot-pourris and to sweet-sleep mixtures.

CORIANDER *(Coriandrum sativum)*: An annual plant that can be sown in its final growing position in late spring. It will also seed itself. For the purpose of aromatics, coriander is mainly grown for its seeds which have a fresh, spicy scent. When dried, they are added to pot-pourris.

Catmint, also called catnip, is adored by cats who will chew it, roll in it and sleep among it. Cats gain great pleasure also from playing with small, mouse-shaped toys stuffed with the dried leaves!

The royal herb garden of King Solomon contained herb and spice plants brought from Egypt – myrrh, frankincense and cinnamon trees, saffron crocuses and calamus.

Bergamot or bee balm is a native of the swampy areas of the United States and Canada, where it was originally used by the Indians to make a refreshing tea. It was first described in a book on American flora in 1569 by Dr Nicholas Monardes of Seville, from whom came its Latin name of *Monarda didyma*.

COTTON LAVENDER *(Santolina chamaecyparissus)*: A perennial with silver-grey leaves, small yellow flowers and a camphor-like scent. Very good in sweet-bags to deter moths.

HYSSOP *(Hyssopus officinalis)*: A perennial that likes to be well cut back in the autumn in order to produce tall, lush growth the following year. Hyssop has small, pointed, shiny dark green leaves and spikes of pink, blue or white flowers that make a good show in late summer and are loved by bees. Its scent is warm and spicy and it can be added to pot-pourris and sweet-bags.

LAVENDER *(Lavandula spica)*: The well-loved aromatic plant with thin, pointed, grey leaves and spikes of purple or mauve flowers. Unusually, it is the flowers that are used most. Put them into pot-pourris, sweet-bags and sweet-sleep mixtures or use them alone to make lavender bottles or lavender bags. There are many different varieties but all have the same fresh, clean scent.

LEMON BALM *(Melissa officinalis)*: A prolific, bushy perennial with tiny white flowers and heart-shaped leaves smelling strongly of lemon and honey. Cut back in mid-summer soon after flowering, the plants will produce a second growth of stems. The dried leaves are used in pot-pourris, sweet-bags and sweet-sleep mixtures. This is another 'bee plant'.

The name hyssop comes from the Greek *azob*, meaning a holy herb, because it was used for cleaning sacred places. The essential oil of hyssop is much prized in perfumery.

In France, cotton lavender is called *garde-robe*, since it was frequently put in clothes chests and linen cupboards to keep away moths. It was also used as a strewing herb on earthen floors.

LEMON VERBENA *(Lippia citriodora)*: A delicate perennial that is best grown in a large pot in a conservatory and put outside during June, July and August. The leaves have a delightful lemon scent and, after roses and lavender, are one of the most common ingredients in all fragrant preparations.

LOVAGE *(Levisticum officinale)*: A tall plant with large, dark leaves and a spicy, celery-like scent that is delicious in the hot sun. The root is occasionally used in pot-pourris and sweet-bags and the root and leaves are used in bath preparations.

Marjoram was once
used as a strewing
herb for scattering on earth-
en or stone-floors. It was
was also rubbed over
wooden floors and furni-
ture to give them a sweet,
clean scent.

MARJORAM (*Origanum vulgare* and others):
Marjoram is an easy-to-grow perennial with a sweet
scent, tiny leaves and pink or mauve flowers which
appear in profusion in mid-summer. Add the dried
leaves and flowers to pot-pourris and other fragrant
mixtures. Whole stalks of flowers and leaves can be
used in dried flower arrangements.

MINT (*Mentha spicata* and others): There are many
different varieties of mint. Those most suitable for
the fragrant garden and for sweet-scented preparations are *M. piperita* (peppermint); *M. spicata* (spearmint); *M. rotundifolia* (apple mint); *M. rotundifolia*
'Variegata' (pineapple mint), which is fruity scented
with beautiful variegated leaves; and *M. citrata* (eau
de Cologne mint).

ROSEMARY (*Rosmarinus officinalis*): A beautiful
evergreen plant with small, spiky leaves, tiny blue
flowers and a sweet, pungent scent. Grow it in a
warm, sheltered spot. The leaves are used in pot-
pourris and many other fragrant preparations.

RUE (*Ruta graveolens*): A tall, strongly aromatic
plant used occasionally in pot-pourris.

SOUTHERNWOOD (*Artemisia abrotanum*): A
perennial that grows to 3 feet (1m) and above, with
feathery green leaves and a lemon-like scent. Use it
in pot-pourris and sweet-bags.

SWEET CICELY (*Myrrhis odorata*): An attractive
plant, wide-spreading and tall, with deeply cut
leaves. For aromatics, use black seeds. They have a
myrrh-like scent and can be added to pot-pourris and
used to make furniture polish.

TANSY (*Tanacetum vulgare*): A tall perennial with
feathery leaves and yellow flowers. It has a bitter
scent and is used in pot-pourris and moth deterrents.

THYME (*Thymus vulgaris* and others): There are so
many varieties of this low-growing fragrant plant
that it is impossible to list them all. The tiny flowers
range in color from white to deep purple and the
leaves can be dark green, bright green or variegated.
Grown together, they make a fragrant display. Bees
love them and they can be added to pot-pourris.

YARROW (*Achillea millefolium*): Often thought of as
a weed, this is nevertheless an attractive plant with
feathery green leaves, heads of tiny white or pink
flowers and a spicy scent.

Spearmint is a favorite
herb of bees. It was
once rubbed on the inside
of new hives so the scent
would attract a swarm of
bees. In England various
types of mints were strewn
in churches and great halls.

Thyme sprigs laid
among clothes keep
away fleas and moths. The
Greeks and Romans
fumigated public places by
burning thyme.

Yarrow has long been
known as a healing
herb, its name coming
from the Greek warrior
Achilles who was able to
cure his soldiers' wounds
with the leaves. In France,
it was once known as *herbe
aux carpentiers* because
carpenters bound it over
cuts sustained while they
were at work.

ROSES

Roses are the most fragrant and possibly the most loved flower of all. They make an enchanting display in a summer garden and, when dried, their scent can last for many years in pot-pourris and other fragrant preparations. A description of all the fragrant roses would itself take up a whole book, so here is a brief list of types and some of the varieties.

GALLICA ROSES: The oldest known rose is *Rosa gallica,* sometimes known as *R. rubra,* the Red Rose, which was grown in temple gardens in ancient Persia in 1200 BC, but which is also thought to be native to Europe. From this came *R. gallica officinalis,* the Apothecary's Rose, which has for centuries been considered the best rose for pot-pourri making, being deep pink with golden stamens and highly fragrant. A sport of *R. gallica* is the beautiful Rosa Mundi (*R. gallica 'Versicolor'*), a deep pink color, splashed with white. Most gallica roses are deep pink, crimson or purple. They need little pruning and their one drawback is that they only bloom in the midsummer months.

Other very fragrant gallica roses include Belle de Crécy; Belle Isis; Charles de Mills; Cardinal de Richelieu; D'Aguesseau; Jenny Duval.

ALBA ROSES: *Rosa alba* is the white rose, which originated in ancient Greece and was introduced to Western Europe in the Middle Ages. Most varieties are highly fragrant but, like the gallica roses, they only flower in midsummer. All tend to grow taller than the gallica roses and most have almost white or pale colored blooms.

Those to look out for include Alba Semi-Plena (one of the oldest garden roses, bearing sprays of milk-white blooms with golden stamens) Maxima; Celestial; Maiden's Blush; Mme Legras de St Germain; Queen of Denmark (Königin von Dänemark).

DAMASK ROSES: *Rosa damascena* is a descendant of *R. gallica,* small and elegant in growth, often with deep pink blooms. Many of the varieties have a sweet rich scent but flowering is only in the midsummer months. These include Mme Hardy Ispahan; Quatre Saisons; La Ville de Bruxelles.

R osa Mundi blooms only at mid-summer. It is said that it was named after Rosamund, the young mistress of Henry II of England, and was planted at Woodstock where they had their secret meetings.

T he damask rose Omar Khayyam was raised from seed from the rose found growing on the grave of the Persian poet Omar Khayyam at Nishapur. It was first cultivated at the Royal Botanic Gardens at Kew in 1884.

Above left: Rosa Mundi
Above right: Mme Hardy
Below: Queen of Denmark

R osa alba and *Rosa rubra* became the symbols respectively of the houses of York and Lancaster during the Wars of the Roses.

T he rose that has been most used for the production of attar of roses and which is by far the best for making pot-pourri is *R. damascena trigintipetala,* or *Kazanlik.* It is deep pink and highly scented.

CENTIFOLIA ROSES: This is the cabbage rose, or the rose of a hundred petals, which has been grown since the sixteenth century. It has also been called the Provence Rose. Cabbage roses grow into fairly large shrubs and flower in midsummer.

Some of the most fragrant are Centifolia (the original cabbage rose, a deep glowing rose-pink); Bullata; Chapeau de Napoleon; Petite de Hollande; Paul Ricault.

Chapeau de Napoleon

HYBRID MUSKS: The musk rose has been grown in gardens since the time of Queen Elizabeth I. It flowers throughout the summer with abundant clusters of blooms. Most of those grown today are hybrids developed during the early part of this century. The most fragrant are Bishop Darlington; Buff Beauty; Cornelia; Felicia; Moonlight; Prosperity.

MOSS ROSES: The moss rose was first developed in the eighteenth century from *R. centifolia* 'Muscosa' and was so called because of the moss-like growth around the bud and flower. Moss roses have a rich, fruity scent and flower in midsummer.

The original moss rose is known as Old Pink Moss and has clear pink blooms. Other highly fragrant varieties include Capitaine John Ingram; Gloire de Mousseux; Little Gem; William Lobb.

BOURBON ROSES: These were developed as a result of crossing damask roses with *R. chinensis* or China roses, which usually have less scent but which flower perpetually. The result is perpetually flowering, richly scented blooms, similar to damask roses. The scent is often reminiscent of raspberries, which is most marked in the almost thornless Zéphirine Drouhin. Other fragrant varieties are Boule de Neige; Bourbon Queen; La Reine Victoria; Louise Odier; Mme Isaac Pereire; Mme Ernst Calvat.

ENGLISH ROSES: English rose-grower David Austin has recently developed a type known as English Roses by means of grafting old-fashioned roses on to perpetually flowering shrubs. Most are compact plants and many are strongly scented.

Try Belle Story; Charles Austin; Charmian; Chianti; Chaucer; Constance Spry; Ellen; Hero; Mary Rose; Sir Walter Raleigh; Warwick Castle; Wife of Bath.

HYBRID PERPETUAL ROSES: These are perpetual

The first Bourbon rose was discovered growing in a garden on the French island of Bourbon in 1817. It was a natural hybrid of a damask rose and a China rose.

Louise Odier

Constance Spry

flowering Victorian roses, rather tall and straight. Many have sweet, rich scents.

Try Baroness Rothschild; Ferdinand Pichard; Hugh Dickson; Baron Girod de l'Ain; Reine des Violettes; Souvenir du Dr Jamain.

RUGOSA ROSES: These are large prickly shrubs which grow in almost any soil. They make good borders and hedges, and bear richly scented blooms.

The most fragrant include Agnes; Conrad Ferdinand Meyer; Fimbriata; Hansa; Lady Curzon; Mrs Anthony Waterer; Nova Zembla; Roseraie de l'Hay.

RAMBLING ROSES: These only flower once in the summer but during that time are covered in massed blooms. They are ideal for growing up trellises and fences but, with the exception of Albertine, are unsuitable for walls.

Some of the most fragrant varieties are Albertine; Alexander Girault; Paul's Himalayan Musk; The Garland.

SWEET BRIAR: *R. rubiginosa* or sweet briar is grown mainly for its sweetly scented foliage which, on mornings when there is a heavy dew, fills the garden with a rich, spicy scent. It can be planted as a hedge or trained along fences or wire. Varieties include Amy Robsart; Greenmantle; Janet's Pride; Lady Penzance; Meg Merrilies.

Reine des Violettes

Most of the present day musk roses were raised in the early twentieth century by the Rev Joseph Pemberton of Essex.

OTHER FRAGRANT FLOWERS

Many other fragrant flowers can scent the garden from early spring to late autumn and can also be dried for use in pot-pourris.

Heliotrope is so called because it follows the rising and setting of the sun. As it opens it faces east and then gradually turns throughout the day to the west. Overnight, it turns back to the east again, ready to open in the morning.

HELIOTROPE (*Heliotropium × hybridum*): Also called cherry pie because that is what its scent resembles. Its leaves are ridged and heart-shaped and the mauve or purple flowers grow in clusters in May and June. 'Royal Marine' is a good variety. Grow as an annual, sowing under glass in winter.

HONEYSUCKLE (*Lonicera periclymenum*): The glorious, trumpet-like flowers on this climbing shrub bloom from spring to autumn. Plant in the spring in rich, well-drained soil.

JASMINE (*Jasminum officinale*): Jasmine is best grown over porches and archways. From early summer to autumn it is covered in tiny white flowers

Lily of the Valley

According to legend, the scent of lilies of the valley growing wild in the woods draws the nightingale to the place where he will find his mate.

with a rich, spicy scent that is stronger in the evening. Plant between autumn and spring in rich, well-drained soil.

JONQUIL (*Narcissus jonquilla*): All narcissi have a sweet fragrance but this one probably has the strongest. The flowers are small and white with orange centers and several grow on each stem. Plant the bulbs in early autumn.

LILAC (*Syringa vulgaris*): In spring, the sweet scent of lilac can fill a garden. The flowers vary from white through to deep purple. Plant in a fertile, sunny spot between autumn and spring and dead-head after flowering.

LILY OF THE VALLEY (*Convallaria majalis*): The tiny, white, bell-like flowers appear among the lush foliage in late spring. Lily of the valley likes damp places in dappled shade.

Above: Honeysuckle
Right: Nicotiana

The mignonette was loved by Victorians and Edwardians for its haunting fragrance.

MIGNONETTE (*Reseda odorata*): An unusual, tall plant with spinach-like leaves and rather plain yellow flower spikes, but nevertheless capable of wafting a strong sweet scent over several yards. Grow as an annual, making successive sowings in spring, at the back of a border in dappled shade.

MOCK ORANGE BLOSSOM (*Philadelphus*): The white, yellow-centered flowers appear on this deciduous shrub in midsummer. They have the delicious scent of true orange blossom and their heads can either be large and single or double in profuse clusters. Plant in winter in well-drained soil in sun or partial shade.

NICOTIANA (*Nicotiana alata*): Sometimes called the tobacco plant, there are several varieties of nicotiana available, in white, cream, yellow and various shades of pink. Treat them as a half-hardy annuals.

ORANGE BLOSSOM (*Choisya ternata*): A lovely evergreen shrub that has heads of starry, white, sweetly scented flowers in late spring and occasionally also in a mild autumn.

Pinks

PEONY (*Paeonia officinalis*): Also known as the Apothecary's Peony, this variety is the traditional deep red. Its scent is very light and the petals are mostly used to give color to a pot-pourri. Plant them in autumn or spring. They may not flower for about two years but after that can last for up to 50 years.

PINK (*Dianthus*): There are many different varieties of these pretty little plants with their long, grey leaves and fragrant, crinkle-edged flowers. The old-fashioned varieties such as Clove Pink and the white Mrs Sinkins tend to be the most fragrant. The plants are perennial and grow best at the front of a border. They flower in the midsummer months.

STOCK (*Matthiola*): *M. maritima* or Virginian stock is the old-fashioned variety with small, four-petalled flowers of various shades of pink and lavender. *M. incana* has sweetly scented flowers growing in thick spikes. *M. bicornis* is a straggly plant with small, pale pink flowers but with a wonderful evening scent. Grow them all as annuals. They flower from early to late summer.

A single flowered variety of peony was introduced to Britain from Europe in the sixteenth century and it was quickly developed into the now well-known Apothecary's Peony which has deep red, double flowers. In Sussex, peony root necklaces were worn to ward off toothache.

Peonies

In Hungary ladies grew a variety of sweet rocket in pots and brought it indoors at night to perfume their bedrooms.

SWEET ROCKET (*Hesperis matronalis*): A tall plant with cabbage-like leaves and long seedpods, this has large heads of white or purple flowers all summer. Grow in rich, moist soil and cut back in the autumn.

SWEET WILLIAM (*Dianthus barbatus*): Favorite cottage garden flowers, these have clove-scented heads of white, pink or mauve flowers in midsummer. Grow them as annuals or short-lived perennials. If they are to flower a second year, shear the plants soon after blooming.

TREE LUPIN (*Lupinus arboreus*): Tree lupins have spikes of yellow flowers, with a strong, honey scent. They flower in midsummer and thrive in sandy soil.

VIOLET (*Viola/cornuta*): Violets grow under bushes and hedges and flower in spring. Once established, the plants will look after themselves.

WALLFLOWER (*Cheiranthus cheiri*): These are among the first fragrant flowers of spring. Their petals are the rich, deep colors of velvet and their scent is strong and sweet. Sow them in the spring and plant out the following autumn for flowering in the following spring.

The tree lupin is a native of California. In 1793 it was introduced into Britain, where it now grows wild on sandy heathland.

Tree lupin

Cornflowers

FLOWERS FOR COLOR

The petals of roses and other fragrant flowers are always used as the base for pot-pourris. For extra color and texture other unscented flowers can be added. The following are all easy to grow annuals:

BORAGE *(Borago officinalis)*
CORNFLOWER *(Centaurea cyanus)*
LARKSPUR *(Delphinium ajacis)*
LOVE-IN-A-MIST *(Nigella damascena)*
MARIGOLD *(Calendula officinalis)*
NASTURTIUM *(Tropaeolum majus)*
PANSY *(Viola × wittrockiana)*
PERIWINKLE *(Vinca major* and *minor)*
POPPY *(Papaver)*
STRAW DAISY *(Helichrysum bracteatum)*

Delphiniums and hydrangeas also make useful additions.

SCENTING A ROOM WITH POT PLANTS

Perhaps the most natural way of scenting a room with the fragrance of flowers is to bring the growing plants indoors. There are two basic types of plant that are most successful — those grown from bulbs and the many of pelargoniums with scented leaves.

Fragrance from Bulbs

Many of the flowers grown from bulbs have delicate, sweet scents that can bring the feeling of spring into any room. When grown in pots, they can either be allowed to flower at their proper time in spring, or they can be grown in such a way that they will flower for occasions such as Christmas and the New Year.

Narcissi, hyacinths, crocuses, lily of the valley, scillas, grape hyacinths and the small irises called *Iris reticulata* can all be grown in pots, but for early flowering the narcissi and hyacinths are the best. For early flowering, you will need to buy bulbs that have been specially prepared. Whichever type you choose, the bulbs must all feel firm when you touch them and they should not be too small.

A container with drainage holes is best for plants that are to be grown outside and then brought indoors. Cover the drainage holes with broken pottery or large pebbles. Bulbs to be forced indoors can be planted in pots or bowls without holes. Use a good quality cutting compost as a growing medium and plant all bulbs in early autumn.

Put a layer of moss or leaf mold in the bottom of the bowl or pot and then cover it with moist compost, the depth depending on the size of your bulbs. For early flowering, when the bowl has been filled, the tips of the bulbs should just reach the surface of the compost which should be about ½ inch (1.5 cm) from the top of the bowl. For spring flowering, the bulbs should be covered with a layer of compost equal to their own height. Lay the bulbs lightly on the compost, close together but not touching, then fill the bowl with more compost, firming it down lightly around the bulbs.

For spring-flowering bulbs, simply put the pots in a sheltered spot in the garden. As soon as the flower buds appear in the spring, bring the pots indoors.

Narcissi were grown in ancient Greece and were probably taken to Western Europe by the Romans. Old English names for them include 'Sweet Nancies', 'None-so-Pretty' and 'Primrose Peerless'.

Grape hyacinths can be blue or white. The white varieties were once called 'Pearls of Spain'. They flower in spring and their scent has been likened to a mixture of honey and grapes.

Iris

24

Put them in a light, warm place, but not in direct sunlight or near a radiator, and water regularly to keep the compost moist but not wet.

Bulbs for Christmas flowering require special care. As soon as they have been planted, they need darkness and a cool temperature in order to allow strong roots to develop. Either put the pot into a black plastic bag and then in a cool, dark cupboard indoors or in the garage or shed, with a constant temperature of between 40° and 45°F (4° and 7°C); or put it in the garden and cover it with a 4 inch (10 cm) layer of peat.

In about eight weeks, or when the shoots are about 1½ inches (4 cm) high, bring the pot, now out of its bag, into a cool room (about 50°F/10°C). For Christmas flowering, this should be no later than the beginning of December. As soon as the leaves have grown and the flower buds are just beginning to appear, put it into the room where it is to flower and treat it as the spring-flowering plants.

SCENTED PELARGONIUMS

The beauty of the scented pelargoniums is all in their leaves. Whereas the flowers tend to be unremarkable, the leaves are magnificent, strongly scented and come in all manner of shapes and textures. Some are soft and furry, others are ridged and deepy cut, and their scents can be anything from mint or peppermint to richly resinous. Well looked after, pelargoniums can last for years, providing indoor scents and foliage all through the year. Touch the leaves for their fragrance to be released and any cuttings can be dried and used in a pot-pourri.

Scented pelargoniums are best bought from a specialist grower who will have a good selection. Some herb farms also stock a wide variety.

Pelargoniums are easy to keep and are very good-tempered. Indoors, they can be placed on a south-facing windowsill for most of the year, although they do prefer to have a little shade in high summer and do not like the atmosphere to be too stuffy or humid. Water them regularly, but only when they feel dry. Feed them with a liquid feed during the summer.

Left to themselves, pelargoniums will grow long and straggly and so it is best to cut your plants into

There is often confusion as to which are geraniums and which pelargoniums. 'Geranium' is the commonly used term for the family of plants known as *Geraniaceae*, which includes pelargoniums. Most of the plants that we know as 'geraniums' are, in fact, pelargoniums.

shape early to encourage side-shoots to grow and make the plants bushy. This cutting back is called 'stopping'. The plants will not flower until 12 weeks after stopping, so it must be done early in the year. It can also be carried on during the flowering season on stems without side shoots.

The central stems that you remove can, from midsummer to early autumn, be used as cuttings. Fill 3½ inch (9 cm) pots with a good garden compost or compost mixed with a little perlite or coarse grit. Then simply stick in your cuttings and firm them down. They should develop roots within four weeks.

Pelargoniums can survive all year indoors, but thrive best if they are given a holiday in the fresh air. You can leave them outside from late spring to autumn, looking after them exactly as you would indoors; alternatively, they can just be placed outside for a few days in every summer month. Some varieties to choose from include:

Pelargoniums are natives of the semi-arid regions of South Africa and they can therefore tolerate hot, dry conditions.

P. denticulata: A tall, strong plant with large, starry, light cut leaves, small pink flowers splashed with crimson and a sweet resinous scent.

P. filicifolium: Thin, finely cut leaves and a resinous scent.

P. 'Lady Mary': Small, tough, lemon-scented leaves and droopy pink flowers.

P. 'Orange King': Large, spiked, coarsely hairy leaves with a strong, oily, orange scent.

P. 'Prince of Orange': Similar to 'Orange King' with a sweet orange scent.

P. 'Lilian Pottinger': Small, soft green leaves and a delicate sweet scent.

P. 'Rodens': Finely cut leaves and a minty scent.

P. fragrans 'Variegata': Variegated leaves with a scent of nutmeg and tiny white flowers.

P. tomentosum: Large, soft, furry leaves, minty scent and tiny pink flowers.

P. 'Lady Plymouth': Variegated leaves and a woody scent.

Essential oil can be extracted from some varieties of scented pelargonium, particularly *P. capitatum*, often called rose geranium, and *P. odoratissimum,* which was once used as an adulterant to rose oil.

DRIED HERBS AND FLOWERS

Herbs and flowers can be enjoyed not only for their fragrance and color in the garden – they can also be dried, a process which will preserve both their appearance and their scent. There are various ways

of doing this and which method you employ depends on the type of plant to be dried and its eventual use.

Herbs for drying will have most aroma if they are harvested just before their flowers are fully out. This also applies to flowers. Petals for drying singly should be picked just as the flower becomes full blown but before they are ready to drop.

Ideally you should harvest all herbs and flowers on a dry day, as soon as the dew has evaporated but before the hot sun has dissipated their scent. Do not cut more than you can deal with at once. Most plants can grow for a few more days without spoiling, but drying too many at a time may hinder the process.

Hanging in bunches

This is suitable for most herbs, either sprigs with leaves only or flowering sprigs such as those of sage or lavender. Flowers with firm stalks, such as helichrysum (straw daisy), which are grown especially for drying, can also be treated in this way, and so too can grasses and seedheads. Divide the cut sprigs into small bunches and fasten them together with fine cotton string, gardeners' twine or raffia. Tie them securely with a slip knot which can be gradually tightened as the stems dry and shrink.

One or two chive, onion or leek flowers look very pretty in a bowl of pot-pourri, and their scents diminish when they are dried. They should be hung up in a different way as their stems are very soft. Suspend a piece of fine wire netting between two points, such as two beams or rafters. Drop the stems through the netting, leaving them to hang down with the flowers resting on top.

Hang the bunches, or place your netting, in a dry, warm, airy room away from direct sunlight. As soon as the leaves and flowers feel crisp take the bunches down, as extra exposure to air and light may cause scents and colors to change or diminish.

One or two well-dried whole flower heads look extremely attractive in a bowl of pot-pourri.

Drying on Frames

This method is most suitable for all single flower heads and for all petals, including rose petals, that are used in pot-pourris both for scent and for decoration. Grasses and flowers with fine, delicate stems also dry well on frames.

From 1 × ½ inch (2.5 × 1.5 cm) wood, make frames about 15 × 10 inches (38 × 25 cm). Stretch muslin over the frames, gluing it down on all sides. The frames can be used singly or stacked one on top of the other if there is limited space.

Roses are just right for harvesting when you can pull off their petals easily. If the petals fall when the rose is shaken, they are too old and their scent has begun to fade. Pick off rose petals in a cool room. They will lose their scent in hot sun.

Place the flowers in a single layer on the muslin. Petals and small flowers can be put close together, as long as they do not overlap. Put the racks in a warm, airy room, preferably one that does not get much sun. After about two days, the petals and smaller flowers will have shrunk considerably so they can be moved closer together to make room for the next batch. Large, fleshy flowers, such as marigolds and pinks, should be turned after three days to ensure even drying. When the flowers are crisp, they are ready to be stored.

Store whole flowers in boxes between layers of tissue paper. Petals can be placed loose in boxes or sealed bags. Keep both in a dark, dry cupboard or drawer.

Flowers and petals dried for their fragrance will keep for up to a year and those for decoration for up to three years.

Drying in Desiccants

Whole flowers for decoration can also be preserved in a desiccant, which will absorb their moisture while leaving their shape and color more or less intact.

The simplest desiccant to use, and also the most easy to obtain, is sand. Used straight from the beach it will be clean enough, but builder's sand needs washing. Put it into a bucket, cover it with water, add a little detergent and swish it about. Leave the sand to settle to the bottom and then pour off as much water as you can. Put the sand on a tray and leave it in the hot sun, or put it into a low oven, for 3-4 hours or until it is quite dry. It is then ready to use.

Silica gel is well-known for soaking up excess moisture and it acts as an effective drying medium for flowers. Its drawbacks are that it is expensive and also that it can often only be bought in crystal form and has to be crushed with a rolling pin between two layers of paper in order to render it fine enough to work efficiently. After use, it must be dried in a low

It has been said that the fragrance of flowers is their music, several substances perfectly blended to make a sweet harmony.

oven until litmus paper turns blue when put to it.

Ordinary household borax is also effective. It is very fine and has to be worked around the petals with a fine paint brush. Adding an equal volume of corn-meal helps to coarsen the texture and make it easier to apply.

Use all three desiccants in the same way. Put a ½ inch (1.5 cm) layer of desiccant into the bottom of a plastic food box. Lay your flower heads on top in a single layer, well spaced apart. Put a 1 inch (2.5 cm) layer of desiccant on top. Repeat the layers until the box is full. Leave in a warm place for 3-4 days, then carefully brush the desiccant from one flower. If it is crisp, then drying is complete.

Remove the desiccant from all the flowers using a paint brush and a spoon.

Store the flowers between layers of tissue paper in boxes until they are needed. They should keep for up to three years.

The writers of many of the old gardening books regarded the scents of flowers and their healing attributes as being their most important qualities. John Parkinson, in his book *Paradisius*, compared fragrant flowers to virtuous men whose qualities live on even after they have died.

Buying Dried Flowers and Herbs

If you are not able to grow and dry your own flowers and herbs, do not despair, for most can be bought dried and ready for use. One or two herbs, such as *Eucalyptus citriodora* or patchouli, are impossible to grow at home and you may not be able to grow sufficient quantities of others.

Small amounts of herbs normally sold for culinary purposes, such as marjoram or rosemary, can be bought off the supermarket shelf and many health food shops carry a wide range. Always buy the best quality and make sure that they have not been kept too long.

Many herb gardens now have their own small shops where you can buy dried herbs and flowers together with plants and seeds.

However, for the best selection of herbs, fragrant flowers and petals go to a herb specialist. If there isn't one in your area, some will supply by mail order. The names and addresses of some of the best can be found on page 149 of this book.

Dried decorative flowers have become very popular in recent years and can be bought in many varieties and colors, from florists, garden centers, craft shops and even supermarkets.

Fragrant Dried Posies

Dried flowers without any scent can easily be rendered fragrant by dropping about 3 drops of oil on each flower head. Use a single oil or a mixture of two. Suitable combinations are: rose and carnation; rose and clove; orange or mandarin oil with orange blossom; lavender and lemon; lavender and rose; lemon and rose.

After adding the oil, seal the flowers in a plastic bag and leave them for 24 hours. This will help to make the scent longer lasting.

Small posies of scented dried flowers can be tied with ribbon and either hung up or placed in a vase.

Making a Fragrant Braid or Plait

Cut 3 pieces of ¼ inch (6 mm) wide colored ribbon 30 inches (75 cm) long. Hold them together and bend over 5 inches (13 cm) at one end to make a loop. Tie a knot in the loop to secure it. Plait the long pieces of ribbon to the end. Tie a knot in the end to secure the braid.

Cut V-shapes in the bottom ends of the ribbon. Trim the loose ribbon at the top to about ¾ inch (2 cm) and neaten by cutting V-shapes.

Take 5 large whole dried flowers that have been scented with oil and stick them on to the ribbon braid with glue. Hold them in place with pins until the glue is dry if necessary. Hang up the braid by the top loop.

DRIED AND BOTTLED INGREDIENTS

Sweet spices, aromatic woods and richly fragrant gums have been used since ancient times to enhance the scents of dried flowers and herbs. Alone, your dried plants can smell wonderful but, apart from the stronger-scented ones such as lavender, their perfume is not robust enough to make a lasting impression on a large room. Exposure to the air can also diminish their fragrance. Spices, woods, gums and essential oils all give subtle undertones to simple flower fragrances without masking them altogether, while some help to 'fix' the scents, making them longer-lasting.

Spices

Spices should be bought whole for most fragrant mixtures, since their aroma begins to change and diminish from the moment they are ground. Only for pomanders, where large quantities of spices are needed, it is best to buy them ready ground.

Whole spices will keep their aromas locked in so they can be bought in fairly large amounts and stored in airtight containers for up to two years. They do not have to be ground to a powder for use, and are in fact best coarsely crushed with a pestle and mortar. Some, such as nutmeg, can be grated.

ALLSPICE: Sometimes called Jamaica pepper and looking like variably-sized black peppercorns, allspice has a soft, sweet aroma that goes well with roses and other delicately scented flowers.

CARDAMOM: Cardamom seeds are sold still in their individual pods, which should be pale green and rounded if the seeds are to be of the best quality. Remove the pods before crushing the seeds to release their sweet, sherbety aroma.

CINNAMON AND CASSIA: The rolled sticks of pale brown, sweetly fragrant bark come from the tree known as Ceylon cinnamon. Cassia is the bark of another variety of cinnamon tree. It has a stronger, sharper scent and looks like small chips of light brown wood bark. These chips are often used whole in fragrant mixtures. Ground cinnamon is a soft brown powder and is often made up of a mixture of cinnamon and cassia.

CLOVES: Sweet-scented cloves are the flower buds of the clove tree (*Eugenia aromatica*), which is native to the East Indies. They are difficult to crush, but really only need a little bruising to release their fragrance. Oranges are studded with cloves to make pomanders and can then be matured in a mixture of ground cloves and other spices.

CORIANDER SEEDS: Tiny, ridged, pale brown coriander seeds have a sweet, fresh scent that is best in mixtures for the living room and kitchen. They are easy to crush and a few go a long way.

JUNIPER: Like allspice, juniper berries have a soft, subtle scent. They are blue-black and slightly soft under their thick skins so that they can be crushed easily. Use them with sweetly scented flowers.

NUTMEG AND MACE: Nutmeg and mace grow together, the mace forming a bright red lacy coating to the shiny brown nutmeg. Once dried, mace becomes bright yellow. It is very hard and needs a good deal of pressure when being crushed in order to release its mild, bitter-sweet scent. Nutmeg has a more robust aroma, making it ideal for scented cushions for a living room.

TONQUIN BEANS: When dried, these are small, wrinkly black beans about the size and shape of an almond. They have a sweet, rich, musty smell and are used both as an underlying scent and as a fixative. They are delicious in spiced bedroom mixtures and grated in pot-pourris and herb sachets.

VANILLA PODS: These are the same long black pods that are used to flavor custard and other desserts. Their aroma is soft, sweet and spicy and is easily imparted to dried flowers and petals. Chop them into ½-1 inch (1-2.5 cm) pieces for adding to pot-pourris and crush the pieces for bags and sachets.

It was the coming of spices from the Far East in the fifteenth and sixteenth centuries that first made pot-pourris popular in Britain and Europe.

Grenada has come to be known as the Nutmeg Isle, although Marco Polo first found nutmeg trees growing in the Far East in 1298.

Vanilla was used by the Aztecs of South America to flavor a cocoa-type drink. When it was taken back to Spain it was described by King Philip as 'that smell of musk and balsam from New Spain'.

Woods

Chips or raspings of sweetly scented woods can often enhance both the appearance and aroma of pot-pourri and sachet mixtures. Buy them in small amounts and store them in airtight jars or sealed packets away from hot sun. The most readily available are sandalwood, sometimes called santal, and sanderswood. Sandalwood is a light brown wood that comes in small chips that look as though they have been swept from the carpenter's floor! It has a gentle, fresh, spicy smell. The chips are added whole to pot-pourris but are too lumpy for bags and sachets.

Sanderswood comes in tiny, red-brown raspings and has a mustier scent. It is small enough to be put into bags and sachets as well as pot-pourris.

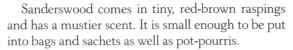

Sandalwood has a sharp, bitter, resinous scent. It comes from India, Australia and the West Indies.

Roots

The roots of some plants are more aromatic than the flowers or leaves and some of these are used in fragrant mixtures.

ORRIS ROOT: This is the most well-known and the most commonly used. It has a strong, violet-like scent and is the best fixative for all pot-pourris and sachet mixtures. It is the root of the small, purple-

flowered *Iris germanica*, which can be grown in most temperate parts of the world and is produced commercially around Siena, in Italy. After the roots have been lifted, they are dried for three years in warehouses after which time they are hard. They are then ground to a creamy-colored powder. Both whole pieces of root and orris powder can be bought from herb specialists, but it is far more convenient to buy the ground variety as its scent does not diminish with time.

If you really wish to be self-sufficient it is possible to grow and dry your own orris. After digging up the roots, peel them and then dry them on racks in a sunny room, turning them at least every two days until they feel hard and sound hollow when tapped. This will take about two months. Wrap them in tissue paper and store them in a drawer for three years. You will find that the scent becomes stronger every year. For use, grind the pieces of root in a coffee grinder kept especially for the purpose.

A corus calamus, or sweet flag, was, in medieval and Tudor times, grown in the fen country of East Anglia as a strewing plant for churches and cathedrals. It was sweetly scented but unfortunately attracted fleas.

CALAMUS ROOT: This is the root of the marsh plant *Acorus calamus*, which is sometimes known as 'sweet flag'. It is an efficient fixative with a mild, sweet, fruity scent, but it is not as readily available as orris root.

BAYBERRY: The bark from the dried roots of this aromatic plant is ground to make a dull brown powder, similar in appearance to ground cinnamon and with a fresh, sweet scent.

VETIVER: Vetiver is a sweetly scented grass with a root which, when dried, smells like a mixture of sweet flowers and aromatic wood. If you can obtain it, this root should be grated into mixtures where it will both scent and act as a fixative. Where it is unobtainable, vetiver oil can be used instead.

The ripe berries of the bayberry give off a wax-like substance when they are boiled and in the United States this was once used to make Christmas candles. Bayberry essence is still a favorite scent for modern festive candles.

Dried Citrus Peel

The dried peels of oranges, lemons and limes are often added to pot-pourris and to sachet mixtures. Orange and lemon rinds can be bought from herb specialists and all can be made very easily at home.

First of all thinly pare the outer rind from the fruits. Put them on a plate or in a bowl and rub in a little orris powder. The rind from 6 large oranges will need 2 tbls orris powder, 6 large lemons need 1½ tbls and 6 limes need 1 tbls.

Put the coated peels on to non-stick baking sheets and put them into the oven, preheated to 300°F/ 150°C. The orange and lemon rinds will take 2½-3 hours to dry and the lime rinds 2¼-2½ hours. They should be hard but not too brittle.

Cool completely and then store in airtight containers. Just before use, coarsely crush or bruise them with a pestle and mortar.

Gums

Gums are the dried, scented resins of certain trees, most of which come from the Far East. They are available in both crystal and powder form. As with the spices, they are far from aromatic in their whole form and must be crushed with a pestle and mortar. Keep them all in airtight jars or packets.

Gum benzoin is often referred to in old manuscripts as gum benjamin or even oil of ben, both of which were easier to write and to say! It comes from the tree *Styrax benzoin*, a native of Java and Sumatra.

GUM BENZOIN: This gum comes in large lumps, or in a grey-brown powder. It has a sweet, musty scent and makes a good fixative.

GUM TRAGACANTH: This is usually only available in a fine, creamy-white powder with a musty, flowery scent.

FRANKINCENSE: Frankincense crystals are small and mostly yellow, with a scattering of light brown pieces mixed in; the powder is a light yellow-brown. The scent is fresh and akin to citrus. The chief use has been to make incense for places of worship.

MYRRH: Myrrh crystals look like brown coffee sugar and the powder is a dull, light brown. The scent is similar to frankincense only more musty. It was once a favorite embalming spice.

Frankincense, also called olibanum, comes from the leaves and bark of several species of tree of the *Boswellia* genus. Its true scent is only revealed when it is warmed.

MUSK: Musk is not a gum at all, but it is similar in appearance and scent and so fits into this category better than any other. It was once produced from the belly of the male musk deer and was reputed to be a great aphrodisiac. Now it is made synthetically. It is a fine, sparkling, white powder with a sweet cloying scent. One of its main uses is in burning perfumes.

Essential Oils and Essences

Every plant has its own 'essence' or 'life force', which can be extracted from flowers, leaves, roots, seeds or berries for use in various fragrant and healing preparations. The best oils are produced commercially by distillation. These are the most expensive but their scents are very true to those of the original plant. Cheaper oils may well have been extracted or expressed by other methods, including chemical ones, and their scents may not be of good quality.

Synthetic oils are produced entirely by chemical means and tend to be cheaper. Their scents can be surprisingly true and they are ideal for making quantities of pot-pourri.

Herbal essences consist of herbal oils mixed with a percentage of alcohol. They can be used in the same way as essential oils.

A few drops of essential oil or essence will go a long way and so should only be added drop by drop, when you are making pot-pourris. A selection can usually be bought in pharmacies and drug stores and most herb specialists carry a large range.

Oils and essences should only be bought in small amounts and they should be stored in small, dark-colored glass bottles in a cool, dark cupboard. This way, they should keep for up to two years.

Alcohol and Spirits

Toilet waters and colognes are made with strong, colorless and odorless alcoholic spirits which have been produced by several processes of distillation. Generally, the stronger the spirit the more easily it will take on the scents of herbs and flowers that are macerated in it. If you are simply adding oils and essences, a weaker spirit will do as well.

In Britain you can buy Polish spirit which is 80 percent alcohol or 140° proof (the equivalent of 160° proof in the United States). This can be bought from off licenses or liquor stores, some stockists of wine-making ingredients and from some pharmacists. If it is not available, vodka, which is usually 46 percent alcohol or 80° proof (92° in the US), can be used instead.

Home distillation was common in the sixteenth, seventeenth and eighteenth centuries. Wines and brandies were distilled several times to make what were known as rectified spirits or spirits of wine.

Recipes

A number of preparations can be made incorporating these dried and bottled ingredients. Home-made furniture polish is especially nice, and several recipes for sweet-smelling polish follow.

Linseed Oil and Vinegar Mixture

4 fl oz (120 ml) linseed oil

— · —

4 fl oz (120 ml) malt vinegar

— · —

40 sweet cicely seeds, chopped and crushed or 1 tsp
each lemon oil and lime oil or 1½ tsp lavender oil
and 4 drops peppermint oil

Pour the oil and vinegar into a jar and seal it tightly.
Shake well to mix, rather like a salad dressing.

Add the sweet cicely or essential oils and shake
again. The sweet cicely mixture should be left for 2
weeks in a warm place and shaken every day. The
mixture with essential oils added can be used
immediately.

To use, impregnate a duster with the mixture and
rub it hard on your furniture or paneling. It will
polish and dust at the same time and there will be no
need to buff afterwards.

Sweet Cicely Furniture Polish

2 oz (50 g) beeswax

— · —

2 oz (50 g) white wax

— · —

4 fl oz (120 ml) turpentine

— · —

60 sweet cicely seeds

Grate the beeswax and put it into a bowl with the
white wax (this usually comes in granules) and the
turpentine. Finely chop and crush the sweet cicely
seeds. Add them to the bowl.

Stand the bowl in a pan with water round it to
come halfway up the sides. Set the pan on a low heat,
without letting the water boil, until both waxes have
melted. Take the pan from the heat.

Pour the mixture into a wide-necked jar and leave
it until it is cold and set. The sweet cicely seeds will
sink to the bottom. Leave the polish for about 1
week. The smell of the turpentine will gradually
diminish. Use like ordinary wax polish, rubbing it on
with one duster and buffing it with another.

The juice of the stems and leaves of lemon balm was used to give a shine to furniture and its sweet lemon-and-honey scent was said to make a man merry and joyful.

The juice extracted from sweet cicely seeds was once used to both polish and scent oak furniture and paneling.

Furniture Cream

4 oz (100 g) beeswax

— • —

16 fl oz (475 ml) turpentine

— • —

8 oz (225 g) lemon balm leaves and stems or 4 oz
(100 g) dried lavender flowers

— • —

1½ pt (900 ml) water

— • —

1 oz (25 g) soap flakes

— • —

1 tsp lemon balm oil or lemon verbena

— • —

oil or 1 tsp lavender oil

Grate the beeswax and put it into a bowl with the turpentine. Put the bowl into a pan with water to come about halfway up. Set the pan on a low heat, without letting the water boil, until the wax has melted.

Meanwhile put the lemon balm or lavender into a pan with the water. Bring them to the boil, cover and simmer for 45 minutes. Strain off the liquid and measure out 10 fl oz (300 ml).

Add the soap flakes to the hot measured liquid and stir well to dissolve. Add the essential oil. Stir the soap mixture into the beeswax mixture and stir to make a light brown cream the consistency of mayonnaise. Pour the cream into jars. Leave it until it is cool and cover it.

Leave the cream for about a week for the smell of the turpentine to diminish. To use, rub the cream into the furniture with one duster and buff with another.

POT-POURRI

If you detect a sweet, pervasive scent as you enter a room, you will probably find a bowl of pot-pourri lending its gentle fragrance to the atmosphere.

Pot-pourri has been made for almost as long as flowers have been cultivated, for it is a way of preserving scent long after the blooming on the plant has ceased. Its fragrance was well known to the ancient Eygptians and Greeks and it has been popular in Britain and Europe since the sixteenth century when spices from the East became more readily available, allowing some to be spared from the kitchen. In the United States, pot-pourri making arrived with the first roses, and by the eighteenth century most country ladies on both sides of the Atlantic were well versed in its arts.

The fragrance of pot-pourri continued to perfume drawing rooms and bedrooms throughout the nineteenth century and the beginning of the twentieth. It was never really forgotten completely but gradually became less familiar after the Second World War. During the middle part of the twentieth century pot-pourri was all but replaced by fresh air sprays, but fortunately the secrets of its making were cherished by a few and now it has become popular again. Not only can you buy it easily but many more people have become interested in making it for themselves.

The name pot-pourri actually means 'rotten pot', and this came from the original method of making it in which rose petals and other fragrant flowers were semi-dried and then fermented in a crock with salt. This type of pot-pourri is now known as a moist pot-pourri.

The more modern kind is known as a dry pot-pourri. It is made with completely dry flowers and petals and it has been more popular in modern times as it is the easiest to make. Also, in recent years, the essential oils needed to give it a lasting scent have been more commonly available.

DRY POT-POURRI

A dry pot-pourri is made from a simple basic mixture of dried flowers and herbs. It is pretty to look at and gently fragrant, is the quickest and easiest of the two types to make and needs no complicated equipment. If you have never made a pot-pourri before, then the dry type is the best to begin with.

Flowers and Herbs

Many English country houses had their own recipes for pot-pourri which were handed down through the generations and never divulged to other households.

Most kinds of fragrant flowers and aromatic herbs can be used to make a dry pot-pourri and you can vary the combination to suit your tastes and the room in which the finished mixture is likely to be placed. One based on roses, for example, will give a sweet fragrance, suitable for a living room or bedroom. A pot-pourri based on herbs such as peppermint or marjoram will have a sharp, clear scent that is best for a kitchen or a downstairs bathroom.

If your finished pot-pourri is fragrant but rather dull in appearance, other flowers can be added purely for color and texture.

Flowers and herbs can be taken from the garden at any time of the year, dried until they are crisp and then stored in sealed containers or bags until you have enough to mix into the final pot-pourri. If you cannot grow enough, you can easily buy them.

Many pot-pourris have been based on the sweet, long-lasting scent of fragrant rose petals. *Elinor Sinclair Rhode*, an authority on herb and fragrances at the beginning of this century, recommended 'the old cabbage roses'.

Other Ingredients and Equipment

In order to make a pot-pouri that will pervade a room with a lasting scent, other ingredients must be added to the dried petals.

Spices, dried citrus peels and fragrant woods give underlying complementary scents and fragrant oils will enhance the overall affect. Fixatives are also added both to preserve the scent of the other ingredients and to add their own underlying aroma.

All the equipment required to make a dry pot-pourri can be found in a modern kitchen. You will need a large bowl for mixing the ingredients, scales, tablespoons and teaspoons, a pestle and mortar (preferably a large one) for crushing spices, roots and peels and a large plastic bag in which to store the final mixture while it is maturing. This may take around six weeks before the raw scent mellows.

A Basic Dry Pot-Pourri

The ingredients will vary within each recipe depending on availability, but the following is a good, basic guide.

1½ pt (900 ml) main fragrance (this will be your main flower such as rose petals or lavender)

— • —

¼-½ pt (150-300 ml) complementary herbs and leaves

— • —

up to 6 tbls spices, crushed

— • —

3-4 tbls dried citrus peel, crushed

— • —

1½ oz (40 g) orris root powder or other fixative

— • —

up to 6 drops scented oils, depending on their strength.

Put the dried flowers and leaves into a large mixing bowl. Crush the spices, citrus peels, gums and roots with a pestle and mortar. Grate others, such as whole orris root or nutmeg, if necessary. Add these to the flowers and herbs. Mix with your fingers.

Add the oils, 1 drop at a time, and mix after each drop. If the scent seems strong enough before you have added all the oils in a recipe, stop. Extra can be added later if necessary, but too much fragrance can spoil the finished pot-pourri.

Put the mixture into a large plastic bag and seal it tightly. Store it for 6 weeks, shaking every other day. The somewhat raw scent which is present at first will gradually mellow and all the separate fragrance will combine together.

Note: Although all the fragrant ingredients for pot-pourri should be mixed together at the same time, flowers for decoration and color can be added at any time as they become available.

All the following recipes should be mixed according to the basic method given here. All ingredients are given in dry weights.

A BASIC BEGINNERS' KIT

When you first start to make pot-pourri there is no need to buy too many ingredients. With just a small selection you will be able to make many different mixtures.

Here is a basic kit which includes flowers which can be grown and prepared at home or which can be bought. Many of the spices and herbs can be found in the kitchen and all are easy to buy.

Flowers and Herbs

Rose petals, plus any other dried fragrant flowers from your own garden such as carnations and pinks, wallflowers, orange blossom, honeysuckle, lilac, nicotiana, peony, stocks, wallflowers (where a recipe says 'rose petals' it can be those alone or a mixture of rose petals with one or more of these flowers); lavender; lemon verbena; bay; rosemary; thyme; marjoram; peppermint; chamomile; marigold; yellow everlasting flowers; dried blue flowers such as hydrangea, cornflower, larkspur or mallow; any other decorative flowers from your garden such as borage, delphinium, pansies, poppy petals, nasturtiums, salvia, zinnias.

Peony petals dry to a deep red and make rose pot-pourris exceptionally attractive. Peonies have for a long time been much-loved country garden plants. They were once thought to shine at night and to protect both sheep and corn from evil.

Spices, Essential Oils and Other Ingredients

Cloves; whole nutmeg; cinnamon sticks; blade mace; allspice berries; vanilla pods; sandalwood chips or powder; dried orange peel; orris root powder. Rose; lavender; peppermint; clove; orange; lemon; cedarwood; carnation; orange blossom essential oils.

ADDITIONS TO THE BASIC KIT

Once you have experienced the delights of making pot-pourri and the wonderful scents that you can create with simple ingredients, you will probably wish to experiment further with extra herbs and spices and more complicated mixtures.

Add the following ingredients to your collection. If you cannot grow and dry them yourself, all the dried flowers and herbs can be bought, apart from geranium leaves. If you cannot obtain them, substitute lemon verbena for the lemon-scented ones and peppermint leaves for the peppermint-scented. If you cannot find an ingredient, do not despair. Pot-

Before the days of plastic bags, large sweet jars or tightly closed earthenware crocks were used for maturing pot-pourri.

pourri recipes are very flexible and adaptable to the availability of ingredients.

Flowers and Herbs

Rose buds; jasmine flowers; woodruff; meadowsweet; lemon thyme; lemon balm; basil; hyssop; costmary (alecost); wormwood; tansy; sweet lemon- and peppermint-scented geranium leaves; hibiscus.

Spices, Essential Oils and Other Ingredients

Cardamom; coriander; juniper; caraway; sweet cicely seeds; tonquin beans; calamus root; lovage root; lemon peel; lime peel; gum benzoin powder or crystals; myrrh powder or crystals; gum tragacanth powder; musk crystals; bayberry powder; sanderswood. Cypress; patchouli; vetiver; ylang ylang; magnolia; honeysuckle; jasmine; French musk; lilac; violet; essential oils.

The recipes should all be mixed according to the method given earlier. All the ingredients are given in dry weights.

A Simple Rose Pot-Pourri

This is an easy one to start with, made with the
simplest ingredients.

6 oz (175 g) rose petals

1 oz (25 g) lemon verbena leaves

1 oz (25 g) bay leaves, crumbled

about 1 oz (25 g) any decorative flowers

6 inch (18 cm) cinnamon stick, crushed

1 oz (25 g) dried orange peel

1 oz (25 g) orris root powder

2 drops rose oil

2 drops lemon oil

Rose and Herb Pot-Pourri

This is a sweet pot-pourri with an underlying fresh
scent of herbs.

4 oz (100 g) rose petals

2 oz (50 g) lavender flowers

1 oz (25 g) marjoram

1 oz (25 g) thyme

1 oz (25 g) rosemary

2 tbls orange peel, crushed

2 tbls cloves, crushed

1 tbls allspice berries, crushed

1 oz (25 g) orris root powder

2 drops rose oil

2 drops orange oil

1 drop clove oil

Lemon Verbena and Lavender Pot-Pourri

This pot-pourri is both sweet and refreshing.

2 oz (50 g) lemon verbena

3 oz (75 g) lavender flowers

1 oz (25 g) peppermint

1 oz (25 g) marjoram

blue flowers for decoration

½ nutmeg, grated

1 tbls blade mace, crushed

1 oz (25 g) orris root powder

1 drop lavender oil

1 drop lemon oil

1 drop orange blossom oil

1 drop peppermint oil

Woody Rose Pot-Pourri

This has a soft, relaxing scent and so is a good
bedroom pot-pourri.

4 oz (100 g) rose petals

2 oz (50 g) dried marigolds

1 oz (25 g) bay leaves, crumbled

1 oz (25 g) sandalwood chips, bruised

1 oz (25 g) dried orange peel

2 drops cedarwood oil

2 drops carnation oil

1 drop clove oil

Oriental Rose Pot-Pourri

4 oz (100 g) rose petals

1 oz (25 g) basil

1 oz (25 g) jasmine flowers

1 oz (25 g) hibiscus flowers

1 tbls cardamom seeds, crushed

2 tbls coriander seeds, crushed

1 oz (25 g) gum benzoin crystals, crushed, or gum benzoin powder

2 drops rose oil

2 drops ylang ylang oil

1 drop patchouli oil

Jasmine and Rosebud Pot-Pourri

Made from whole flowers, this has a particularly attractive appearance and an old-fashioned scent reminiscent of Edwardian drawing rooms and bedrooms.

1 oz (25 g) jasmine flowers

2 oz (50 g) rose buds

½ oz (15 g) peony petals

1 oz (25 g) calamus root, crushed or grated

2 tbls sanderswood

1 tbls cloves, crushed

1 tbls juniper berries, crushed

1 tbls myrrh crystals, crushed, or myrrh powder

2 drops jasmine oil

1 drop rose oil

Rose, Hibiscus and Peony Pot-Pourri

This is a particularly attractive mixture of pinks and reds, with a sweet, gently spiced fragrance, suitable for living room or bedroom.

4 oz (100 g) rose petals

2 oz (50 g) hibiscus flowers

1 oz (25 g) peony petals

2 tbls dried lime peel, crushed

½ oz (15 g) dried orange peel, crushed

6 tonquin beans, grated or crushed

2 vanilla pods, chopped and crushed

1 tbls sweet cicely seeds, crushed

½ oz (15 g) bayberry powder

1 oz (25 g) orris root powder

2 drops rose oil

3 drops French musk oil

Melissa Pot-Pourri

Melissa is the old name for lemon balm, which has a sweet, lemon scent. If you grow it yourself, use whole dried leaves as they look more attractive and also increase the bulk. This is a sweet pot-pourri with a refreshing hint of lemon, suitable for the living room or bedroom.

1 oz (25 g) lemon balm leaves

2 oz (50 g) rose petals

½ oz (15 g) basil

½ oz (15 g) thyme

½ oz (15 g) poppy petals

½ oz (15 g) dried orange peel, crushed

1 oz (25 g) sandalwood chips

1 oz (25 g) calamus root, crushed

1 tbls juniper berries, crushed

1 tbls blade mace, crushed

1 tbls broken cassia bark

1 oz (25 g) gum tragacanth powder

2 drops lemon balm oil

2 drops rose oil

2 drops orange oil

Country Lane Pot-Pourri

The sweet, herby scent of this pot-pourri is reminiscent of summer days spent in the country where the scents of wild flowers and freshly cut hay mingle with those of herbs and flowers from adjoining gardens. It has an unusual green and yellow color and is suitable for kitchen, living room, bathroom or dining room.

1 oz (25 g) chamomile

1 oz (25 g) woodruff

1 oz (25 g) meadowsweet

1 oz (25 g) marjoram

1 oz (25 g) rue

1 oz (25 g) hyssop

1½ oz (40 g) yellow everlasting flowers

1 oz (25 g) dried lemon peel, crushed

1 nutmeg, grated

3 drops honeysuckle oil

2 drops lilac oil

Kitchen Mix

This mixture will help keep your kitchen smelling fresh at the same time as keeping away flies. It looks particularly attractive in a wooden bowl decorated with the dried flowers of chives or onions, which should have lost all their scent.

1 oz (25 g) costmary

1 oz (25 g) tansy

1 oz (25 g) lavender

1 oz (25 g) peppermint

½ oz (15 g) lemon-scented geranium leaves

½ oz (15 g) mint-scented geranium leaves

½ oz (15 g) bay leaves

½ oz (15 g) rosemary

1 oz (25 g) orris root powder

½ oz (15 g) lovage root, crushed

3 drops rosemary oil

2 drops lemon oil

2 drops patchouli oil

MOIST POT-POURRI

A pot-pourri made by the moist method has a sweet, rich scent that will last for five years or more. This was the original pot-pourri, made over a number of months in earthenware crocks kept in cool cellars and work rooms. It was made by the Greeks and ancient Egyptians who buried jars of fermenting petals in the ground, and was popular in British, European and American country houses until the end of the nineteenth century.

A moist pot-pourri is more complicated to make than a dry one and you must have your own supply of fresh rose petals, but the extra effort is well worth while for the wonderful lasting scent that you will achieve in the end.

In the sixteenth century, houses were small with no damp courses and often with mud floors. The overall smell was musty and stale. Large bowls of pot-pourri were said to 'sweeten the ayre'.

Ingredients

Whereas some dry pot-pourris can be made with a mixture of other herbs and flowers, roses always form the base of a moist pot-pourri. Pick them on a dry day just before they are full blown and carry them in a basket, a cardboard box or even in an improvised bag made from a cotton sweater or cardigan – never in a plastic bag, which will make them sweat.

As soon as you get the roses home, pull off the petals and scatter them on drying racks, preferably in a single layer. They can be put on deeper if you do not have enough racks, provided that you turn them several times a day. Leave them in a warm, dry, airy place until they are the texture of soft leather, still pliable and half dried out. This will take about two days. Other fragrant garden flowers can be dried in the same way and mixed with the rose petals, but roses must always predominate.

The ancient Greeks and Egyptians preserved rose petals by burying them in crocks in the ground.

Once the pot-pourri has been started off, other similarly prepared petals can be added every two to seven days throughout the summer. Freshly grated lemon and orange rind can also be mixed in.

The other main ingredient that you will need is pure salt, which has no iodisers added or any other ingredients to make it 'flow'; a pure sea salt is perfect.

As the fermentation process itself produces such a strong, lasting scent, there will be no need for fixatives.

Spices are nearly always included in a moist pot-pourri. Ground spices can be added with the salt and at this point it is best to add only the sweeter, milder types such as cinnamon and cloves that always enhance the scent of roses. You can, however, add nothing to the roses until fermentation is complete. After fermentation, when you know how strong the scent of your pot-pourri is going to be, you can add larger amounts of freshly crushed spices to produce the required end result. Barks and roots are always added fermentation.

A large, glazed earthenware crock has always been the main essential piece of equipment. Use a bread crock or even a large mixing bowl.

You will also need a measuring cup or jug and a heavy plate or wooden chopping board to act as a weight on top of the petal and salt mixture.

Meadowsweet has a sweet, heady scent when it is dried. It was the favorite strewing herb of Elizabeth I and was used in great halls and living quarters in the summer.

Proportions for Initial Mixing

1½ pt (900 ml) rose petals (or rose petals with a small proportion of other fragrant flowers)

— · —

½ pt (300 ml) pure salt

— · —

grated rind ½ orange or lemon (optional)

— · —

½ tbls ground cinnamon (optional)

— · —

½ tbls ground cloves (optional)

Basic Method for Moist Pot-pourri

Mix the petals, grated peels, spices and salt in your crock bowl. Weight them down and leave them in a cool, dry place to ferment. If you are continually adding petals to the mixture, add salt and spices with them in the correct proportions and stir after each addition.

Alecost has a fresh scent reminiscent of spearmint. Its large, gray-green leaves have often been used to scent bedrooms and linen cupboards. It was also a popular strewing herb.

After the last flowers have been added, leave the crock for two weeks and then check the contents. If a sweet-smelling liquid has collected in the bottom of the crock, pour it off. It makes an excellent addition to your bath water. By this time, the mixture should be beginning to shrink.

Replace the weight and leave the mixture for six weeks, checking every week. By the end of the process, you should be left with a dry, layered, sweet-smelling cake of petals. If you have used only red and

pink roses, this will be pink. A mixture of colors plus spices will make a dull, brown color.

Break up the cake of petals and add dried herbs and more spices. Mix well and seal the mixture in a plastic bag. Store it in a dark place for four weeks before using.

In this type of pot-pourri only about 4 tbls herbs to every 8 oz (225 g) of dried petals is needed, plus about 3 oz (75 g) spices and ½-1 oz (15-25 g) orris root powder. Small amounts of woods and barks can also be added. Oils are not needed since the scent of the petals is so strong.

A pot-pourri made by the moist method will not be as attractive as a dry pot-pourri, but its scent will be far stronger and more lasting.

The spices and herbs that you mix with your petals depend very much on personal taste and availability. Here are three totally different examples, each made with petals that have been fermented with grated lemon rind, ground cloves and cinnamon.

The mixtures are for 8 oz (225 g) dry petals and all herbs are dried.

Cottage Parlor Pot-Pourri

1 tbls lemon thyme

— • —

1 tbls hyssop

— • —

2 tbls marjoram

— • —

1 oz (25 g) cloves, crushed

— • —

½ oz (15 g) cinnamon sticks, crushed

— • —

½ oz (15 g) orris root

Tansy will keep flies away from the kitchen, whether it is hung up in bunches or mixed into a pot-pourri.

Country Manor Pot-Pourri

4 tbls lavender

— • —

2 tbls rosemary

— • —

1 oz (25 g) lemon peel

— • —

1 oz (25 g) juniper berries, crushed

— • —

1 oz (25 g) allspice berries, crushed

— • —

½ oz (15 g) orris root powder

Woodruff has always been prized for its scent of new-mown hay. It was hung up in bunches to relieve stuffiness in closed rooms, laid among linen and used as a filling for cushions and pillows.

Arabian Nights Pot-Pourri

4 tbls basil

— • —

1 oz (25 g) cassia bark

— • —

1 oz (25 g) coriander seeds, crushed

— • —

1 oz (25 g) cardamom seeds, crushed

— • —

1 oz (25 g) orange peel

— • —

½ oz (15 g) orris root powder

CARING FOR YOUR POT-POURRI

When a pot-pourri has matured for the required period of time it can be put into a container ready to fulfill its purpose of perfuming a room.

Pot-pourris perform best in rooms that are warm, light and airy. Never put them in direct sunlight as this will dissipate the scents too quickly and will also fade the colors of a dry pot-pourri. A constantly humid atmosphere will stop the fragrance pervading a room and may in time alter the scent.

After about three months, the scent of a dry pot-pourri will start to diminish. You can revive it by adding small amounts of freshly prepared ingredients, such as crushed spices or newly dried petals and herbs, and extra essential oils, keeping them the same as those in the original recipe.

Dry pot-pourris rarely keep their true scents for longer than two years. The original fragrance changes and becomes unbalanced. Burn them on an open fire and fill the bowls with new mixtures.

The scent of a well-made moist pot-pourri can last for ten or more years. However, should you find it diminishes after a time, add extra spices and revive the rose petals with a little brandy or eau de cologne and one or two drops of rose oil.

CONTAINERS

Dry pot-pourris have a double beauty – their scent and their appearance – and so they are usually displayed in open bowls. These can be made of pottery, glass, wood, silver or silver plate – in fact anything but plastic. You can now buy specially made pot-pourri bowls with patterns cut in the sides which enable more of the scent to be released.

One of the beauties of having pot-pourri in a room is to have the scent waft out at you when you open the door. However, should you wish to preserve the scent of a dry pot-pourri for longer than usual, the mixture can be kept in a lidded container of which the lid is removed only when the room is occupied.

Pot-pourris made by the moist method are not particularly attractive and so they are mostly kept in opaque lidded jars with narrow tops, such as ginger jars. They can also be kept in covered boxes and caskets and in perforated pottery pomanders.

In the seventeenth and eighteenth centuries, moist pot-pourris were kept in lidded china pot-pourri jars which were uncovered for a short time each day. Some had two lids, one airtight to preserve the fragrance, and another perforated one underneath which enabled only a little fragrance to escape. For a stronger perfume, both were removed.

Eighteenth- and nineteenth-century pot-pourri jars were highly elaborate and often decorated with flowers and birds. One in the Victoria and Albert Museum is made of marble and ormolu and dates from 1810.

Of course it is possible to create pretty pot-pourri containers from objects found in any home. Several ideas are outlined below.

This basket of flowers is one of the easiest containers to make and is suitable for either dry or moist pot-pourris. Buy a small, lidded wicker basket. If the bottom of the basket is also made of wicker, cover it with a fitting sheet of brown paper to prevent powdered spices from falling out. Stick dried flowers on the lid. Even with the lid closed the pot-pourri will give a slight scent. For a stronger aroma, wedge it open.

Shells of many sizes make excellent containers and perhaps are most suitable for a bathroom. You will need exotic shells of varying sizes and some nylon net. Make bags from the net, fill them with pot-pourri and put them into the shells with no seams showing.

Old recipes for moist pot-pourris contained other fragrant flowers in the fermenting pot besides roses. They included jasmine, pinks, orange flowers and violets.

Queen Catherine of Braganza's favorite pot-pourri was a simple mixture of rose petals, rosemary, thyme, lavender, orange peel and cloves.

Scent Jar for a Moist Pot-Pourri

7–10 fl oz (200–300 ml), attractively
shaped, clear glass jar
— • —
pressed flowers
— • —
1 lightly beaten egg white
— • —
net material
— • —
satin finished material
— • —
shirring elastic
— • —
matching ¼ inch (6 mm) ribbon.

Make sure that the jar is clean and paint egg white on the inside. Stick your pressed flowers, facing outwards, on the sides of the jar. Leave them to dry. Fill the jar with moist pot-pourri.

For the covering, measure across the top of the jar and add 4 inches (10 cm) for the overlap and an extra ½ inch (1.5 cm) for the turning. When you have the final measurement, cut a circle of this diameter from both the net and the satin finished material. Cut a 1½ inch (4 cm) diameter circle from the center of the satin finished circle only. Press under and machine stitch ¼ inch (6 mm) all round the edge of circles.

Make ¼ inch (6 mm) snips at regular intervals around the central hole in the satin finished material. Press back ¼ inch (6 mm) round the edge. Lay the net circle on top of the satin finished circle. Pin them together round the hole in center. Machine stitch.

Run a line of shirring elastic ⅝ inch (16 mm) in from the edge of the hole. Try the cover on the jar and pull up the elastic to fit if necessary. Secure the loose ends tightly. Tie the ribbon over the elastic.

The salt once most commonly used in old pot-pourris was bay salt. This was a coarse sea salt originally produced in France in the Bay of Bourgneuf and later all along the Atlantic coasts of France, Northern Spain and Portugal. It was the most effective salt for preserving meats.

Lemon verbena leaves have always been a popular ingredient in pot-pourri as their sweet lemon scent lasts for a long time after drying.

Sweet Sleep Pillows, Sacks and Sachets

Small cushions or sacks full of pot-pourri or a specially made mixture of herbs and spices will contribute a delightful fragrance to any room. Put them among the cushions on chairs and sofas – they will release their scent whenever they are touched, besides making pleasant back and head rests. Place them on a dressing table, lay them on the bed or tuck them between the pillows in the bedroom. Hang them up with twine or ribbon in any room.

Besides simple cushions and bags, you can also make other bedroom accessories such as tissue box covers or nightdress cases so that they can be filled with sachets of fragrant mixtures.

The best material to use for all filled bags and sachets is a pure cotton, or a polycotton mix with a large percentage of cotton. This type of material 'breathes' well and so releases the most scent. Buy thicker weaves with colors matching your furniture and other cushions for the living room and lighter materials with delicate colors and patterns for the bedroom. Sacks for the kitchen can match the curtains or you can make them from hessian or burlap or from a bright linen tea towel. You can also make knitted sachets from soft, pure wool.

All covers for cushions and sachets should ideally be washable and so it is best that your mixtures are not put directly inside them. Instead, make sachets of muslin or calico, just a fraction smaller all round, that can be removed easily when the cover is soiled.

You can leave your sachets plain or decorate them with ribbons, embroidery or patchwork, whichever suits your fancy. For the fillings, you can use any of the pot-pourris mentioned on the previous pages, or the sweet-sleep mixtures following these patterns.

Rectangular Sachet

16 inches (40 cm) of 45 or 36 inch (115 or 90 cm)
wide material

— • —

9 inches (24 cm) of 45 or 36 inch (115 or 90 cm) wide
muslin or calico

— • —

2 yd (1.8 m) of ¼ inch (6 mm) wide matching ribbon

Cut 2 pieces of the main material, each 11¼ × 9¼ inches (29 × 24 cm).

For the frill, cut strips of material 3¼ inches (8 cm) wide to a total length of 70 inches (178 cm).

Cut 2 pieces of muslin or calico each 10¾ × 8¾ inches (27 × 22 cm).

First make the frill. Join the two strips together on the short sides with ¼ inch (6 mm) seams to make a circle of material. Press the seams open. Press the strip in half lengthwise, wrong sides together. Make two lines of gathering, one ⅝ inch (16 mm) from the edge and the other ¼ inch (6 mm).

Clothes were once hung on lavender and rosemary bushes to dry.

Pin ribbon round the outside of one rectangle of material, 1⅝ inch (4 cm) from the edge, folding it neatly at the corners. Trim it to fit. Baste, and machine stitch it into place down both edges.

Carefully draw up the gathers in the frill until the frill will fit the outside measurement of your rectangles of material.

With raw edges together and the frill facing inwards, pin and then baste the frill to the right side of the decorated rectangle of material, making sure that the gathers are even. Place the second rectangle on top, with right sides together. Pin or baste it into place. Machine stitch round the outside, making a ⅝ inch (16 mm) seam and leaving a gap of 4 inches (10 cm) on one short side. Trim the seam and turn the cover right way out. Press it.

Cut the remaining ribbon into two equal pieces. Make them into bows. Sew the bows on to opposite corners of the ribbon decoration on the cover. Cut the ends into V-shapes to prevent them from fraying. Place the two pieces of muslin or calico together and sew the edges, making a ⅝ inch (16 mm) seam and leaving a gap of 4 inches (10 cm) at one short edge. Trim the seam and turn the bag.

Fill the bag with your chosen mixture and sew up the open end. Put the bag inside the cover. Sew up the open end of the cover by hand, making the stitches as invisible as possible.

Round Sachet

24 inches (61 cm) of 45 or 36 inch (115 or 90 cm) wide material

— • —

12 inches (30 cm) of 45 or 36 inch (115 or 90 cm) wide muslin or calico

— • —

1⅓ yd (1.2 m) of ¼ inch (6 mm) wide ribbon

Using a pair of compasses, draw and cut out 2 circles of the main material, 11½ inches (29 cm) in diameter. For the frill, cut strips of material 5¼ inches (13 cm) wide to measure 70 inches (178 cm) in total length. Cut out 2 circles of muslin or calico 10½ inch (27 cm) in diameter.

Make the frill, attach it to 1 circle of material and join the 2 circles together using the same method as for the rectangular sachet.

Trim the seam. Turn the case right way out and press it. On the underside of the cushion, pin out of the way the unstitched length of material at the gap and, with the right side up, machine stitch round the edge of the case, close to the frill.

Make and fill the muslin or calico bag as above. Put it inside the material case. Sew up the gap by hand. Cut the ribbon into 4 equal-sized pieces and make them into bows. Sew them around the edge of the sachet with equal gaps between.

The name lavender comes from the Latin 'lavare' which means 'to wash'. It was given to the plant because of the ancient household custom of putting lavender sprigs and bags among newly washed clothes and bed linen. In the twelfth century a washerwoman was actualy called a 'lavenderess'.

Making a Press Fastened Back

Allow 2 inches (5 cm) extra length of material and make the back of the sachet in 2 pieces, one the length by half the width of the front piece and the other the length by half the width plus 2 inches (5 cm) for the overlap.

For example, if the front measures 10 × 8 inches (25 × 20 cm), for the back you will need 1 piece 10 × 4 inches (25 × 10 cm) and another 10 × 6 inches (25 × 15 cm).

Bind one long edge on each of the back pieces using crossways strips or bias binding. When sewing the two sides of the sachet together, put the narrower piece against the front part first with right sides together, and then the wider piece so that when the sachet is stitched and turned, the wider part will lap under the narrow part. When the sachet is finished, secure the two parts together with press studs or snaps.

Lavender was one of the plants taken by the Pilgrim Fathers to America.

Making a Tied Back

This is suitable for all flat sachets. After cutting out your 2 main pieces of material, cut 1 in half crossways. These will make the underside of the sachet.

To bind the edges you will need either 1 inch (2.5 cm) wide strips cut on the bias from the same material or matching bias binding, both the same length as the 2 cut edges. For the ties, you will need lengths of bias strip, bias binding or matching ¼ inch (6 mm) ribbon 10 inches (25 cm) long. For the circular or rectangular sachets above, 3 ties will be sufficient so you will need 6 of these lengths.

To make ties from the bias strips, press in ¼ inch (6 mm) on each long side and also on one end. Fold them in half, press and machine stitch. If you are using bias binding, fold in 1 raw edge, fold the binding in half and stitch. If you are using the ribbon, fold in 1 end twice and sew it by hand.

Who will buy my sweet lavender? was a familiar London street cry in the eighteenth century.

Bind the cut edges of the material, catching in the raw ends of the ties on the wrong side and making sure that they are in the same places on each piece of material. Sew the 2 sides of the sachet together in the normal way, making sure that the 2 bound edges of the underside butt against each other and that the ties are opposite one another.

Heart-shaped Sachet

12 inches (30 cm) of 45 or 36 inch (115 or 90 cm)
wide material

— • —

12 inches (30 cm) of 45 or 36 inch (115 or 90 cm)
wide muslin or calico

— • —

70 inches (178 cm) of 1 inch (2.5 cm) wide ribbon

— • —

20 inches (50 cm) of ⅜ inch (1 cm) wide ribbon

Using the pattern on page 146, cut out a heart shape in stiff paper to use as a template.

Cut 2 heart shapes from the main material. Cut 2 from the muslin and then trim off ¼ inch (6 mm) all round the muslin only.

Pin 2 strips of ⅜ inch (1 cm) ribbon diagonally across a piece of the main material. Baste and then machine them into place on both sides of the ribbon.

Run a line of gathering stitches ¼ inch (6 mm) from one edge of the wider ribbon. Gather up the ribbon to fit round the outside of the heart shape. Starting at one side, pin the ribbon to the edge of the decorated piece of material, with right sides together, with the line of gathers ⅝ inch (16 mm) from the edge of the material and with the main part of the ribbon facing inwards. Fold the ends back where they meet, so the raw edges will not show. Baste the ribbon in place.

Put the second piece of material on top with right sides together. Pin, baste and machine stitch a ⅝ inch (16 mm) seam. Turn the sachet and press it. On the right side, machine stitch around the edge of the sachet in the same way as for the round sachet above, taking care not to include the opening.

Make and fill the muslin or calico bag as for the other two sachets. Put it into the sachet and sew up the edge by hand. Sew the ends of the ribbon together by hand at the join.

Dolly Bag

With different materials and fillings, a dolly bag can be used to perfume any part of the house.

12 inches (30 cm) of 45 or 36 inches (115 or 90 cm) wide material.

— • —

6 ½ inches (16.5 cm) of 45 or 36 inch (115 or 90 cm) wide muslin

— • —

18 inches (45 cm) of ¼ inch (6 mm) wide ribbon, plus extra if a loop for hanging is required

From the main material cut a rectangle 12 × 17 inches (30 × 43 cm) and a circle 6¼ inches (16 cm) in diameter. From the muslin, cut a rectangle 6½ × 16 inches (17 × 41 cm) and a circle 6 inches (15 cm) in diameter.

For the outside bag, join the 12 inch (30 cm) sides of the rectangular piece of material, right sides together and with a ⅝ inch (16 mm) seam. Press the seam open.

With right sides together, insert the round piece of material in one end of the first piece, to make a bag with a round base. Pin the circle into place, baste and stitch with a ⅝ inch (16 mm) seam. Trim the seam, turn the bag and press it.

Neaten the top edge of the bag, either with a zig-zag stitch or by turning in and machine stitching ¼ inch (6 mm). Turn in and press 4 inches (10 cm) at the top of the bag. Make gathering stitches through all thicknesses 2¼ inches and 2½ inches (5.5 cm and 6 cm) down from the top fold.

For the muslin bag, join the 6½ inch (16.5 cm) sides in a ⅝ inch (16 mm) seam. Press the seam open. With right sides together, insert the round base in one end of the muslin tube. Trim the seam and turn the bag. Press in 1½ inches (4 cm) at the top of the bag. Make rows of gathering stitches ¾ inch (2 cm) and 1 inch (2.5 cm) down from the top fold. Fill the bag with the appropriate fragrant mixture. Pull up the gathers and tie the ends to secure them.

Put the muslin bag inside the outer bag. Pull up the gathers on the outer bag and tie them to secure them. Tie a ribbon around the gathers.

Knitted Sachets

Gently scented knitted sachets are warm, soft and comforting and they make delightful head or neck rests in a living room or garden room.

Use a thick, pure wool, such as Aran knitting wool or a handspun wool, in a natural color. Different designs with naturally dyed or other soft-colored wools can be knitted in, depending on the filling – for example, a pattern in pale lavender color for a lavender-scented sachet, or a rose design for one that contains a pot-pourri.

For the perfect headrest, knit 2 rectangles 12 × 8 inches (30 × 20 cm). Cut out 2 pieces of muslin 7 × 5 inches (18 × 13 cm). Join the pieces of muslin together with a ⅝ inch (16 mm) seam, leaving a gap for turning at one side. Trim the seam and turn it to the inside. Fill the bag with lavender or pot-pourri and sew up the gap.

Sew the knitted pieces together on 3 sides. Fill the resulting bag with washed sheep's wool or with toy stuffing, placing the muslin bag in the center. Sew up the remaining side.

Other sizes of knitted sachet can also be made. One 6 inches (15 cm) square is good for tucking under your cheek when lying down. Filled with lavender, it makes a comforting cure for headaches.

Scented Nightdress Case

This case will scent both your bedroom and your nightwear. It is made so the filling can be removed for easy washing.

24 inches (61 cm) of 45 inch (115 cm) wide material

— • —

26 inches (60 cm) of 36 inch (90 cm) wide material

— • —

matching lining: as main material

— • —

30 inches (76 cm) of 45 or 36 inch (115 or 90 cm) wide muslin or calico

— • —

4 yd (3.65m) bias binding

— • —

2 large, 4 small press studs or snaps

Cut a piece of material 21½ inches (54.5 cm) long and 14½ inches (37 cm) wide and round off all the corners. Cut another piece 12½ inches (31.5 cm) long and 14½ inches (37 cm) wide and round off the two bottom corners. Cut a long strip, going across the material, 39½ x 2 inches (1 x 5 cm). (If you have only 36 inch (90 cm) wide material, cut 2 strips each 30 inches (76 cm) long, join them in a ¼ inch (6 mm) seam and press the seam open.) Cut lining pieces in the same way.

Cut 2 pieces of muslin or calico 21½ x 14½ inches (54.5 x 37 cm) and 2 pieces 12½ x 14½ inches (31.5 x 37 cm). These can be cut across the material.

Pin the pieces of main material to the pieces of lining with wrong sides together. Machine stitch all round the edges of the two large pieces with a ⅛ inch (3 mm) seam, leaving a gap of 6 inches (15 cm) in the center of the top of each piece. There is no need to machine stitch round the strip.

Make a snip ¼ inch (6 mm) in the lining on either side of this gap on the smaller piece of material and pin the part between the snips down and out of the way. Bind the top edge, with both lining and main material in the binding on either side of the gap and the main material only between the snips. Now bind the raw edge of the lining between the snips. This will make an opening with two bound edges.

Bind one short end of the strip, taking in both lining and material. With wrong sides together, pin the strip to the remaining three sides of the smaller

piece of material before reaching the end, trim the end of the strip to fit and bind it. Stitch the strip to the main piece in a ⅛ inch (6mm) seam.

Pin the strip to the larger piece of material in the same way and machine with a ⅛ inch (3 mm) seam.

Bind all the edges, making the gap in the top of the larger piece of material in the same way as that on the smaller piece.

Fold over the top of the large piece to make the flap and sew on the large press studs or snaps to secure it, making sure that the stitches go through one layer of material only. Sew 2 small press studs or snaps on each of the bound openings.

Put the matching pieces of muslin or calico together and sew round in a ⅝ inch (16 mm) seam, leaving a gap of 6 inches (15 cm). Trim the seam and turn the resulting flat bags. Fill the muslin bags with pot-pourri or a sweet-sleep mixture, taking care not to overfill them and to keep them flat. Between them they should take about 12-14 oz (350-400 g) of mixture.

Insert the muslin or calico bags through the bound openings in the edges of the case and rearrange the fillings to keep the case flat.

Tissue Box Cover

A tissue box with a cover is so much more attractive than the plain box and with a scented filling in the top it will gently perfume a bedroom or living room, releasing a mild fragrance as each tissue is pulled out. Use a sweet-sleep mixture or a pot-pourri for the filling, depending on where the box is to be put.

You will need only a very small amount of material, the exact amount depending on the size and shape of the box. Measure the top of the box. Add ½ inch (1.5 cm) to both the length and width measurements and cut out a piece of material with those dimensions. Do the same with the sides. Measure from the top edge of the front of the box, round underneath to the top edge of the back. Add ½ inch (1.5 cm). Cut out a piece of material of that length by the width of the box plus ½ inch (1.5 cm). Cut out lining pieces the same size.

For the inner bags, cut 2 pieces of muslin the same measurements as the piece for the top of the box.

Cut each piece in half crossways.

To find the length of bias binding required, add together the outside measurement of the top of the box, two box lengths and twice the outside measurement of the box sides.

You will also need 8 ¼ inch (6 mm) press studs or snaps and 2 self-covering buttons.

Cut in half crossways the main piece and lining piece for the top. Lay the main pieces on the lining pieces, wrong sides together. Machine all round the edge with a ⅛ inch (3 mm) seam, leaving a gap in the center of the cut edge of 3 inches (7.5 cm) in the smallest cover and 6 inches (15 cm) in the larger ones. Make a ¼ inch (6 mm) snip in the lining at either side of the gap.

Make 2 loops of bias binding the size of the buttons plus ¼ inch (6 mm). Pin one loop to the lining side of half of the top, midway between the gap and the outer edge. Pin the other to the other half of the top, diagonally opposite the first. Bind the inner edges of the top, taking care not to include the lining between the snips. Bind the lining between the snips separately.

Pin the side and wrap around pieces to their respective linings, wrong sides together. Stitch them all round in a ⅛ inch (3 mm) seam. With wrong sides together, pin a top piece to each end of the wrap around. Stitch in a ⅛ inch (3 mm) seam. Bind seam.

Bind the top edge of each side piece. With the top edge of the side pieces level with the bound edge of the top, pin and baste the side pieces into position. Stitch with a ⅛ inch (3 mm) seam. Bind the seam.

Cover the buttons and sew them in position, taking care to sew only through one layer of material. Sew on 2 press studs or snaps at each side to fasten the bound edge of the top to the bound edge of the lining. Sew press studs or snaps to fasten the top pieces to the side pieces.

Put 2 pieces of muslin together and sew ⅜ inch (1 cm) seam all round, leaving a gap of 3 inches (7.5 cm) on one side, to make a flat bag. Turn the bag. Do the same with the other 2 pieces of muslin. Fill the bags with your chosen mixture and sew up the openings. Insert the muslin bags in the top of the cover. Fasten them in with the press studs or snaps.

Sunshine and Butterflies Mobile

12 inches (30 cm) of 45 or 36 inch (115 or 90 cm)
wide yellow cotton material

— • —

1 piece each orange, red, purple, blue and green
cotton about 12 x 6 inches (30 x 15 cm). (Other
colors or patterned materials may be used according
to preference and availability.)

— • —

Two 7 inch (18 cm) diameter circles of muslin

— • —

5 pieces of muslin 3 x 1¼ inches (7.5 by 3 cm)

— • —

small amount of soft toy stuffing

— • —

nylon thread

Cut 2 circles of the yellow material 7¼ inches (18
cm) in diameter. Either directly on the folded yellow
material, or on to paper to make a template, draw 4
concentric circles, 4¾ inches, 6 inches, 6½ inches
and 10 inches (12 cm, 15 cm, 17 cm and 25 cm) in
diameter. Divide the 3 center circles into 16 sections,
taking the dividing lines to the edge of the 6½ inch
(17 cm) diameter circle (see diagram page 147).
Mark a point on the outer circle opposite every other
dividing line of the inner circles. Draw from the
other lines of the inner circles to these points to make
the outside points of the sun. Cut round the outside
of these points. Cut out the inner circle.

Using the template on page 147, cut out 10
butterfly shapes (2 pieces of each color). Put the 2
pointed circles together, right sides inwards. Pin,
baste and machine stitch a ¼ inch (6 mm) seam all
round the other pointed edge. Trim the points and
snip into the corners. Turn and press the points.
Edge stitch all round the outer pointed edge.
Machine stitch ⅝ inch (16 mm) in from the inner
edge. Snip into the seam allowance at ½ inch (1.5
cm) intervals around the inner edge.

Pin the points to the right side of one of the yellow
circles with the points facing inwards. Baste and
machine stitch in place. Place the second circle on
top with right side inwards. Pin, baste and machine

stitch a ⅝ inch (16 mm) seam, leaving a gap of about 3 inches (7.5 cm). Trim the seam and turn the circles. Press. Topstitch around the edge of the circle, leaving the underside of the gap free. Press in the machine allowance on the free edge. Edge-stitch round fold.

Join the 2 round pieces of muslin together in a ⅝ inch (16 mm) seam, leaving a gap of about 3 inches (7.5 mm). Trim and turn. Fill the bag with mixture and sew up the gap. Put the muslin bag inside the sunshine. Sew up the gap by hand.

Join the 2 pieces of each butterfly together with a ¼ inch (6 mm) seam, leaving a gap along one wing for turning. Clip into the curves. Turn and press.

Make 5 small muslin bags, fill them with the pot-pourri or sweet-sleep mixture and sew ends.

Fill the butterflies with the toy filling, placing the small muslin bags inside. Sew up the gaps by hand.

Attach the butterflies at varying heights to the sunshine with the nylon thread. Attach a piece of nylon thread to the top of the sunshine for hanging.

SWEET-SLEEP MIXTURES

A sachet containing sweet-sleep mixture will make a bedroom smell delightful, besides helping you to relax and sleep well at the end of the day.

Hops are renowned for their narcotic effect and form the base of many sweet-sleep mixtures. They have been used alone to fill large pillows, but this way they tend to give a rather cloying scent and so are best mixed with other fragrances.

Bay, rosemary and *Eucalyptus citriodora* will sharpen the scent and lemon balm (itself an extremely soothing herb), lemon verbena and mint will make it fresher. Woodruff, agrimony, marjoram and southernwood will give sweet undertones.

Lavender was much favored by Victorian and Edwardian ladies as an herb both to relax and to ease

George III could not sleep without his hop pillow.

In the sixteenth century, mattresses were filled with sweet grasses and herbs such as ladies' bedstraw (*Galium odorata*).

Fresh Hop Sleep Mixture

2 oz (50 g) hops

— • —

½ oz (15 g) *Eucalyptus citriodora*

— • —

1 oz (25 g) lemon balm

— • —

1 oz (25 g) orris root powder

— • —

3 drops lemon balm oil

Sweet Hop Mixture

2 oz (50 g) hops

— • —

2 oz (50 g) agrimony

— • —

1 oz (25 g) meadowsweet

— • —

1 oz (25 g) woodruff

— • —

1 oz (25 g) cloves, crushed

— • —

1 oz (25 g) dried orange peel, crushed

— • —

1 oz (25 g) orris root powder

— • —

2 drops sweet orange oil

headaches. Its scent clashes with that of hops, so mix it with other herbs and also with rose petals.

Sweet-sleep mixtures can be put into small sachets, nightdress cases, tissue box covers and nursery mobiles.

A large pillow stuffed completely with an herbal mixture can be overpowering and also uncomfortable! It is best to fill it mainly with an ordinary pillow stuffing, placing in the center a large muslin bag filled with your chosen sweet-sleep mixture. If the end of the pillow is sewn up by hand, the muslin bag can easily be replaced when the scent begins to diminish. The scent of a well-made mixture should, however, last for up to four years.

Make a sweet-sleep mixture in the same way as a pot-pourri, crushing the larger herbs before mixing.

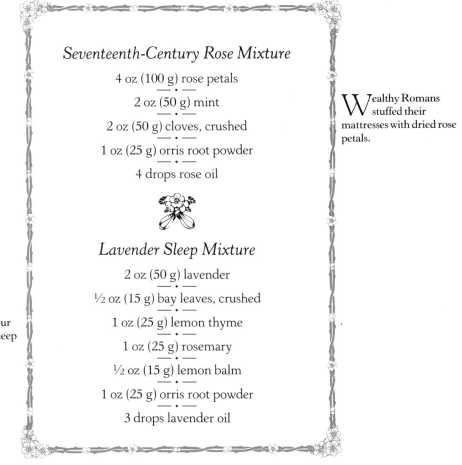

Seventeenth-Century Rose Mixture

4 oz (100 g) rose petals

2 oz (50 g) mint

2 oz (50 g) cloves, crushed

1 oz (25 g) orris root powder

4 drops rose oil

Wealthy Romans stuffed their mattresses with dried rose petals.

Lavender Sleep Mixture

2 oz (50 g) lavender

½ oz (15 g) bay leaves, crushed

1 oz (25 g) lemon thyme

1 oz (25 g) rosemary

½ oz (15 g) lemon balm

1 oz (25 g) orris root powder

3 drops lavender oil

Rosemary under your head while you sleep was once said to keep away evil spirits and nightmares.

SCENTING LINENS AND CLOTHES

To have freshly scented, insect-free bed linen and sweetly perfumed clothes was almost essential in the times when houses smelled musty and the streets worse. Even as conditions improved, it was a sign of a well-run household if the sheets smelled pleasant and there were no moth holes in the blankets. Even now, every woman knows that clothes that waft a hint of fragrance are a joy to put on, but many people find the smell of modern detergents and fabric softeners unpleasant. The softer, subtler scent of herbs and spices is less cloying.

Fresh and dried herbs smell sweet, and some have properties which discourage moths and other insects. Lavender, costmary and rose petals have always been the most popular scents; moth and insect repelling herbs include rosemary, wormwood, southernwood, tansy, woodruff and rue. Cloves, caraway seeds, coriander seeds and orris root powder were often mixed with them.

Sweet-bags can be made in all shapes and sizes and, as with the large sachets, the best material to use is cotton. You can place them in drawers and between piled clothes and linens in cupboards, hang them from hooks in cupboards and wardrobes or hook them over coat hangers. Fill them with a favorite pot-pourri mixture or with one of the special sweet-bag mixtures given below.

Clothes can also be scented by adding herbal decoctions to the final rinsing water after washing and also by sprinkling them with scented water before they are ironed.

The inside of drawers and cupboards can be rubbed with a sweet oil, and wallpaper can be scented and used as drawer liners.

LAVENDER BOTTLES

Lavender has a fresh, clean scent and has been a favorite 'closet herb' for centuries. It deters moths (and fleas!). As well as being refreshing and relaxing, the scent of lavender is both strong and lasting, which means that the herb can be used alone without oils or fixatives.

When lavender stalks are bent over the heads, they make their own containers or 'bottles'. For the simplest, all you need is thread to tie them with, but you can also make them pretty with ribbons and strips of material.

To make one lavender bottle, cut 20 lavender stalks when the flowers have just opened, making them all the same length and as long as possible. Strip off all the leaves and side shoots.

Lay the stalks on a table with the tips of the flower heads at the same level. Pick them up and then carefully tie them at the base of the flower heads with green cotton.

Holding the bunch of lavender upside down, fold the stems down one by one and arrange them evenly around the heads like a cage, keeping the flowers inside. Tie the stems together again about 1 inch (2.5 cm) away from the tops of the flower heads to make a long, even, oval shape rather than a short, dumpy one. You will find that different types of lavender make slightly different shaped 'bottles' – those with long, slender flower heads will be long and thin and those with shorter, fuller heads wider and dumpier.

Make another tie at the bottom of the stems, this time making a loop for hanging. Trim the stems to equal lengths. Hang the 'bottles' in a warm, dry, airy place for 3 weeks until the flowers are dry.

For hanging in a wardrobe the bottles can be left completely plain, but they are much nicer if you tie a bow of narrow ribbon over the thread at the base of the 'cage' and make a loop of ribbon for hanging.

You can also weave ribbon or strips of material that matches your decor in and out of the 'cage', basket fashion. Attach a safety pin to the end to make the process easier.

Lavender bottles can be hung in the wardrobe or cupboard, on the dressing table or on the end of the bed. They can also be put in drawers.

Elizabeth I favored sandalwood for her linens and clothes, King Henry of France violet powder and Edward IV a mixture of anise and orris powder.

A mixture of orris root and spices was at one time blown with bellows on made beds and also on to clothes once they had been put on.

LAVENDER BAGS

These tiny lavender bags can be scattered in drawers or hung from coat hangers and from the slats in an airing cupboard. Or simply place the bags amongst folded clothes and linens. Make them from any scraps of cotton material and tie them with ribbons.

Cut 2 pieces of material 3 × 5 inches (7.5 × 13 cm). Put them together, right sides inwards, and sew round two long sides and one short side with a ⅜ inch (1 cm) seam. Trim the corners. Turn and press the bag. Turn in 1½ inches (4 cm) at the top and press. Run a gathering stitch all round the bag, 1¼ inches (3.5 cm) from the top.

Fill the bag with dried lavender. Pull up the gathers and tie them. Tie ribbon round the gathers, making a small bow if the bag is to be placed in a drawer, or a long tie for hanging in a cupboard or wardrobe.

As soon as costmary was introduced into England in the sixteenth century it was appreciated for its fresh, moth-deterring scent and was often mixed with lavender for placing in the linen cupboard.

SCENTED DRAWER LINERS

Scented drawer liners are easy to make. You will need some wallpaper with an attractive pattern and without too shiny a surface. The cheaper wallpapers that are not 'wipe-clean' or 'steam resistant' tend to absorb the most scent, but you can also use wallpaper to match your bedroom decor. Then, if there is curtain material to match, sweet-bags for placing amongst your clothes can be made from the same pattern. Cut the wallpaper to the size of the drawers that it is to line.

Make up your favorite sweet-bag mixture. Put a few drops of a matching essential oil on a cotton ball or a pad of cotton wool. If lavender were the prominent scent, for example, you would use a lavender oil. Rub the pad of cotton wool on the underside of the wallpaper.

Sprinkle sweet-bag mixture quite thickly over the pattern side of one piece of liner. Lay another on top and sprinkle that with the mixture. Continue until all the liners are together. Roll them up and seal them in a plastic bag. Leave them for 6 weeks.

Brush away all the sweet-bag mixture from the paper before lining your drawers, but reserve it as it can still be used as a filling for sachets.

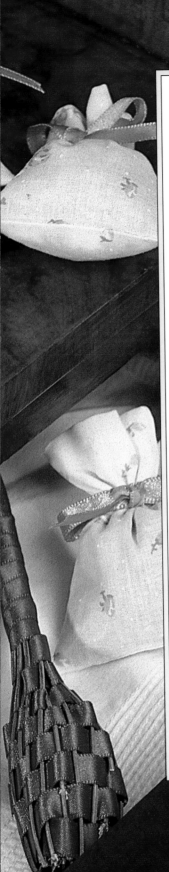

Queen Isabella's Sweet Bags

2 oz (50 g) rose petals

1 oz (25 g) coriander seeds, crushed

½ oz (15 g) orris root powder

1½ oz (40 g) calamus root pieces, crushed

Old Colonial Sweet Bags

2 oz (50 g) rose petals

1 oz (25 g) lavender

1 oz (25 g) hyssop

2 tbls lime peel, crushed

2 tbls blade mace, crushed

2 drops patchouli oil

2 drops vetiver oil

Sweet and Fresh Moth Deterrent

1 oz (25 g) costmary

1 oz (25 g) lavender

½ oz (15 g) lemon verbena

1 oz (25 g) cloves, crushed

½ oz (15 g) orris root powder

Herbal Sweet Bags

1 oz (25 g) rosemary

1 oz (25 g) peppermint

1 oz (25 g) lemon balm

½ oz (15 g) lemon verbena

1 oz (25 g) cinnamon sticks, crushed

1 oz (25 g) cloves, crushed

1 oz (25 g) caraway seeds, crushed

1 oz (25 g) orris root powder

1 oz (25 g) dried lemon peel, crushed

THREE SIMPLE SWEET-BAGS

These three small, simple sachets are perfect for placing among clothes and linens in drawers and cupboards. They can all be made from small scraps of material or from material that matches your drawer liners.

Round Sweet-Bag

2 circles of material, each 6¼ inches
(16 cm) in diameter

— • —

2 circles of muslin each 5¾ inches
(15 cm) in diameter

— • —

18 inches (46 cm) of gathered broderie anglaise
ribbon ½ or ¾ inches (1.5 or 2 cm) wide

Pin the ribbon ⅝ inch (16 mm) from the edge of one of the material circles, on the right side and with the outer edge of the ribbon facing inwards. Baste. Put the 2 circles of material together, right sides facing. Join them with a ⅝ inch (16 mm) seam, leaving a gap of 3 inches (7.5 cm). Trim the seam and snip into it. Turn and press the bag.

Join the 2 pieces of muslin together with a ⅝ inch (16 mm) seam, leaving a 3 inch (7.5 cm) gap. Trim and snip into the seam. Fill the muslin bag with a sweet-bag mixture. Sew up the gap. Put the muslin bag into the material bag. Sew up the gap by hand.

Square Sweet-Bag

2 pieces of material 5¼ inches (13.5 cm) square

— • —

2 pieces of muslin 4¾ inches (12 cm) square

— • —

18 inches (46 cm) of gathered broderie anglaise
ribbon ½ or ¾ inches (1.5 or 2 cm) wide

Make as for the round sweet-bag, clipping the corners when turning the corners.

Besides cotton, sweet-bags can be made from a double layer of nylon netting, the color of which should be chosen to match the color of the filling mixture. The net can also be embroidered with the motif of the main flower or herb contained in the mixture.

In the eighteenth century quilted and perfumed liners were made for linen baskets. A strongly scented sweet-bag mixture was sandwiched between two sheets of thin cotton, with two lengths of taffeta on the outside. The materials were secured together round the edges and then quilted.

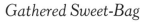

Gathered Sweet-Bag

1 circle of material 8½ inches (21.5 cm) in diameter

— • —

30 inches (76 cm) of gathered broderie anglaise
ribbon ½ or ¾ inches (1.5 or 2 cm) wide

— • —

8 inches (20 cm) of ¼ inch (6 mm) wide ribbon

Press in ¼ inch (6 mm) all round the edge of the circle. Pin the broderie anglaise to the pressed-in edge of the circle, overlapping on the underside. Machine stitch it into place. Run a gathering stitch 1½ inches (3 cm) in from the neatened edge of the circle of material and another ¼ inch (6 mm) below it. Pull up these gathers so you begin to make a small bag. Fill with a sweet-bag mixture. Pull the gathers up tight and secure them. Tie a ribbon round the gathers.

Caraway seeds, when crushed and put into sweet-bag mixtures, give a refreshing and long-lasting camphor-like scent.

FILLED COAT HANGER COVERS

1 wooden 16 inch (41 cm) coat hanger

— • —

6½ inch (16.5 cm) piece of 36 or 45 inch
(90 or 115 cm) wide cotton

— • —

2 yd (2m) of ¾ inch (2 cm) wide ribbon to match

In the center of the material, mark out, on the wrong
side, a rectangle 6½ x 24 inches (16.5 x 61 cm). Draw
a semicircle with a 3¼ inch (8 cm) radius at either
end. Cut out the piece you have drawn. Make a line
of gathering stitches all round the outside of the
material, ¾ inch (2 cm) in from the edge. Bind the
edge with the ribbon. Pull up the gathering stitches
so the cover fits the coat hanger. Put in the coat
hanger. Fill the covering with a sweet-bag mixture.
Distribute the gathers evenly and pin each side of the
cover together along the line of gathers. Either sew
the sides together by hand with a small back stitch or
by machine, using a zipper foot.

SCENTED RINSES AND CASTING BOTTLES

There are two further ways of scenting clothes and linens; you can either add decoctions of herbs to the final rinsing water after hand washing with a detergent or washing compound without too strong a scent, or you can sprinkle them with scented waters as they are being ironed.

Lavender, Rose and Herb Rinse

1 oz (25 g) dried lavender flowers

— • —

1 oz (25 g) dried rose petals

— • —

½ oz (15 g) dried sage

— • —

½ oz (15 g) dried rosemary

— • —

3½ pt (2 liters) water

— • —

4 drops oil of cloves

Put herbs into a pan with the water. Bring them to boil. Cover and simmer for 1 hour. Leave decoction to stand overnight. Strain it and add clove oil.

For use, rinse the clothes once in cold water and squeeze them dry. Fill a sink with cold water. Add the decoction. Put in the clothes. Leave them to soak for 10 minutes. If possible, dry naturally.

Spiced Rose Water for a Casting Bottle

½ pt (300 ml) rose water

— • —

4 drops lavender oil

— • —

1 tbls cloves, crushed

— • —

2 cinnamon sticks, broken into small pieces

— • —

1 chip nutmeg

Put the rose water into a clear bottle. Add the lavender oil. Put in the spices. Seal the bottle and place it in a warm place or on a sunny windowsill for 2 weeks. Strain the liquid and return it to the bottle.

Louis XIV of France had his shirts soaked in a decoction of cloves, marjoram, lavender and rosemary simmered in rose water.

Underclothes and bed linen were often washed in lavender or rosemary water for a clean, fresh scent.

POMANDERS AND SPICE ROPES

Pomanders and spice ropes are a pleasure to make and a joy to have around the house. Not only are they attractive and sweetly scented but they also bring the spicy aromas of bygone days into living rooms and bedrooms. In wardrobes and drawers they act as moth repellents and in the kitchen they keep away flies; they will freshen a bathroom and sweeten the air in a sickroom. Make them for yourself, for gifts and for decorations for special occasions.

A pomander in medieval times consisted of a small, perforated container, filled with an amber-colored ball of scented gums and resins such as ambergris, gum benzoin, civet and musk.

Pomanders were often hung round the neck or round the waist and their sweet scents were inhaled to ward off infection or simply to mask unpleasant smells. Some were used to induce sleep and others were hung in rooms to sweeten a musty atmosphere.

The fruit pomander was devised in the sixteenth century. It was cheaper, easier to make and has remained popular ever since.

By the early twentieth century, both types of pomander could be found in country houses. Most were by this time made from china or porcelain; they were round with perforations in the top and could be hung from ribbons or stood on a table and filled with a pot-pourri made by the moist method (page 58) or with a sponge soaked in perfume.

Spice ropes are a modern invention which have become increasingly popular in recent years. They are made by plaiting lengths of thick wool and tying on bags filled with pot-pourri or spice, together with dried flowers or cinnamon sticks. They are hung from wooden rings and can be placed in any room of the house, depending on their scent.

Spice Ropes and Pot-Pourri Ropes

Spice and pot-pourri ropes can be hung on doors and walls and also inside wardrobes or over coat-hangers. Use pot-pourri ropes in the living room and bedroom, spice ropes in the kitchen, hall and bathroom.

9 × 50 inch (127 cm) lengths tapestry
wool, plus extra for tying

— • —

1 wooden curtain ring 1½-2 inches (4-5 cm) in
diameter, metal loop for hooks removed

— • —

4 circles of material 6 inches (15 cm) in diameter,
cut with pinking shears if possible

— • —

small amount of any type of pot-pourri,
or the spice mixture following

— • —

essential oils of the same type that were
used in the pot-pourri mix

— • —

3 dried flowers or small pieces cinnamon stick

The name pomander comes from the French *pomme d'ambre*, meaning 'apple of amber'. This came about because of the color of the scented mixture and the round shape of many of the perforated containers. The pomander belonging to Elizabeth I was apple-shaped and divided into hinged segments.

Loop the lengths of wool through the curtain ring so that all the ends are together. Tie the lengths just below the ring with a small piece of wool.

Divide the lengths into 3 sections of 6 and braid or plait them. Tie them with another small piece of wool about 3½ inches (9 cm) from the bottom and trim the ends to make them level.

Run a gathering stitch all round each circle of material, 1½ inches (4 cm) in from the edge. Draw up the gathers to make a small bag. Fill it with the pot-pourri or spice mixture. Add 2 drops of essential oil to the pot-pourri. Pull the gathers tight and tie.

Tie a piece of wool round the neck of each bag. Using the same wool, tie the bags on to the rope so that they are equally spaced. Tie another piece of wool around the neck of each bag in a small bow.

Tie a piece of wool to each dried flower or piece of cinnamon stick. Tie these on to rope between bags.

The scent of the rope should remain strong for about a year. You can revive it by adding 2 drops of essential oil to the mixture in the bags.

Spice Mixture for a Spice Rope

Makes enough for 2 ropes

2 tbls ground cinnamon

— · —

2 tbls ground cloves

— · —

2 tbls ground nutmeg

— · —

2 tbls sandalwood powder

— · —

4 drops clove oil

— · —

4 drops orange oil

In Tudor times, cheap pomanders were made by mixing a small amount of spices with ordinary garden earth.

Making a Fruit Pomander

Fruit pomanders can be made with oranges, lemons, limes and even apples. All the fruit should be of the best quality, even in size and shape and unblemished. Choose thin-skinned citrus fruits as these dry quicker than thick-skinned, and shiny, green apples.

Make up an orris root mixture before you start to stick the cloves into your chosen fruit. For each fruit you will need about 3 tbsps of the mixture, which can be made in large batches and will keep in a sealed plastic bag for about 3 months.

To make, simply mix 2 oz (50 g) powdered orris root together with 2 oz (50 g) ground spices.

All spices should be ground. Those sold for culinary purposes are suitable. It is quite possible to use cinnamon alone, but you can experiment with various combinations. A mixture of cinnamon and cloves, for example, would work well with apples, and the freshness of ground cardamom goes well with limes. Other spices you can try are nutmeg, allspice and coriander.

The use of scented oils will produce a stronger perfume in the finished pomander. They can be used to soak the cloves, added to the spice mixture or painted on after the curing process.

The mustier scented oils such as clove, carnation or musk go well with oranges and apples. Citrus, lemon or lime oil add a refreshing touch to the scent of lemon and lime pomanders.

To impregnate cloves with oil, put 1 oz (25 g) cloves into a plastic container with ½ tbls scented oil. Cover tightly, shake well and leave for 24 hours, shaking every so often. Use as below. Pomanders made with impregnated cloves generally need a longer drying period.

To scent the spice mixture, add 1 drop oil per 1 tbls mixture before coating the fruit. Mix well.

To add the oil after curing, brush all the orris mixture from the fruit. Paint the fruit with the oil, using a small paint brush and making sure all the cloves are coated. The pomander must be left to dry for a further period after this.

You will also need about 1 oz (25 g) cloves per fruit. These must be of good quality, with large heads and strong stems. You can also add essential oils.

Cardinal Wolsey was the first recorded person to have favored a fruit pomander. He carried an orange studded with cloves to ward off the plague.

Cloves and clove oil have long been used in the making of cosmetics and perfumes. They were ingredients of the original pomanders and in Tudor times ground cloves were used to scent gloves.

Use a thin bodkin, knitting needle or skewer to pierce the holes for the cloves; a bowl for mixing the spices; a small brown paper bag for each pomander; a pastry brush for brushing the spices off when the fruits have dried.

Gently squeeze citrus fruits with your hands. This will make the skin softer and the cloves easier to insert. If you are going to hang the pomander with ribbon, remember that it is best to leave a cross shape round the fruit free of cloves so that the ribbon will fit snugly round the finished pomander. Either mark this with your first lines of cloves or stick a cross of tape around the fruit before you begin putting in the cloves.

Cinnamon is one of the sweet spices. It is a member of the laurel family and its scent is thought to have aphrodisiac qualities.

The cloves should be spaced with the width of a clove head around each one. The pomander will shrink considerably during the curing process, thus drawing the cloves together. If they are too close in the first instance the pomander may warp or crack.

If the skin of the fruit is very soft, the cloves may push in easily. If not, pierce small holes with your bodkin or other implement first. This saves time (and a sore thumb!) if you are making several pomanders together.

Pat the orris root, spice and scented oil mixture into the pomander, covering every part of the surface. Put the pomander into a paper bag together with any remaining spice mixture. Leave it in a cool, dry, airy room until it is dry, checking every other day to make sure that the pomander remains coated with the mixture and turning it. If, after the first few days, the orris root mixture on the pomander still looks damp, brush it off with the pastry brush and coat the pomander with some of the drier mixture in the bag.

Drying the fruit can take from about 1 to 3 months, depending on the type of fruit and factors such as the atmosphere of the room and even the weather. When the fruits are completely dry they will sound hollow when they are tapped and they will have become smaller. All the fruit will darken to a brownish color during the drying process.

Before decorating the pomanders brush off all the orris root mixture with a pastry brush. Wrap 2 strips of ribbon round the cross left in the pomander by the cloves or tape. Sew them together at the top. Make a bow or several loops of ribbon for decoration at the top and sew them on. Sew on a 6 inch (15 cm) loop for hanging. Dried flowers or bows of very narrow ribbon can be stuck or sewn on the first ribbons.

A pomander with no space for a ribbon may be put into a colored net bag secured at the top with ribbon. This makes the pomander ideal for scenting drawers of clothes and linens. Dried flowers may be stuck or sewn on to the bag or loops of very narrow ribbon tied through the net. If the bag is tight-fitting, a tassel of long narrow ribbons may be tied from the base.

Four or more similarly scented pomanders can fill a decorative bowl to be used as a fragrant centerpiece

Early pomanders were not necessarily round – death's heads or flower shapes were also popular. They were hung round the neck and also from a chain or belt round the waist. Later, the mixture was also put into small, decorated boxes known as *casolettes*. They were carried by doctors at the time of the Great Plague.

In the seventeenth century, pounded precious stones were added to pomander mixtures which were then known as Goa stones.

for a table. Dried flowers can be stuck directly on to these pomanders for a pretty effect.

Pomanders can be decorated for special occasions. Limes, for example, can be adorned with red, green, gold and silver and hung in rows along old beams or from the mantelpiece at Christmas. As a christening present, use soft blues and pinks. Other events for which you can choose suitable embellishments include Mother's Day, weddings and anniversaries.

Green Apple Pomander

Use Granny Smith or Golden Delicious apples.

For each apple:

1 oz (25 g) cloves

— • —

1½ tbls orris root powder

— • —

½ tbls ground allspice

— • —

½ tbls ground cloves

— • —

½ tbls ground cinnamon

— • —

3 drops oil of cloves

Make as above, mixing the oil of cloves into the spice mixture before curing the apples.

An Elizabethan Pomander

Put this mixture into a small perforated pot. Its scent is spicy and fresh and, when smelled at close quarters, very head-clearing. It is easy to see why a pomander such as this was thought to keep away infection.

The sunflower oil is a modern addition, but it helps to keep the ball of spices together and brings out their scent.

1 oz (25 g) gum benzoin

— • —

1 oz (25 g) frankincense

— • —

1 oz (25 g) myrrh

— • —

½ oz (15 g) ground cloves

— • —

½ oz (15 g) ground cinnamon

— • —

½ oz (15 g) sandalwood powder

— • —

6 tbls rose water

— • —

4 tbls sunflower oil

Put the gum and spices into a bowl; add half the rose water and half the sunflower oil and mix well. Add the rest, 1 tbls at a time, until the spices can be brought together in a firm ball.

In ancient Persia, cloves and clove oil were used in the making of love potions.

Cinnamon sticks are made from the fragant inner bark of the cinnamon tree, dried in the sun to make it curl into quills.

Renewing the Scent of a Pomander

The scent of a well-made pomander should last for at least one year. By this time you may well have many more that you have made to take its place. The scent, however, can be revived by rubbing the pomander with 2 tbls of the orris root mixture to which you have added 2 drops of scented oil. Seal it in a plastic bag and leave it for 2 weeks.

The following recipes demonstrate the different ingredients that can be used and the different ways of using scented oils. Once you are accustomed to making pomanders, however, you will come to know your own favorite methods and scents and the recipes can be varied within the basic guidelines, as you wish.

Fresh Lemon Pomander

For each lemon:

1 oz (25 g) cloves

1½ tbls orris root powder

1 tbls ground cinnamon

1 tbls ground allspice

1 tbls ground nutmeg

approximately 1 tbls lemon oil

Make the pomander as above. After it has dried, paint it with lemon oil. Put it into a paper bag, without the spice mixture, for a further week, or until it is completely dry.

Sweet Lime Pomander

For each lime:

1 oz (25 g) cloves

1½ tbls orris root powder

½ tbls ground coriander

½ tbls ground cardamom

½ tbls ground cinnamon

3 drops lime oil

Make as above, adding the oil to the spice mixture.

Classic Orange Pomander

For each large orange:

1 oz (25 g) cloves

½ tbls orange flower oil

1½ tbls orris root powder

1 tbls ground cinnamon

½ tbls ground cloves

Make as above, soaking the cloves in the orange flower oil for 24 hours before sticking them into the orange.

NATURAL FRAGRANCE FOR THE BATHROOM

Scented waters for cleansing your face and hands; waters, oils and aromatic vinegars for adding to the bath; naturally fragrant soap and shampoo – all will help to give a touch of luxury to the time that you spend in the bathroom.

In the days when there was no running water, jugs and bowls of water would be put out at night time ready for a morning wash. Sweetly scented flower petals were sometimes floated on the water, or infusions and decoctions of herbs and flowers were added to it to prevent it from going stale overnight. In Victorian and Edwardian houses, guests were often given a jug full of marjoram water which had a refreshing, sweet fragrance.

Herbal scents in the bath can relax or revive you, relieve tiredness and aching limbs and help the circulation. They are also just wonderful to luxuriate in. Add them by making scented waters and oils, aromatic vinegars which will soften your skin and bags of herbs and oatmeal that you can hold under a running tap. If your skin needs a cleansing scrub, make mitts of herbs and cornmeal. Although you can buy many different kinds of scented soap, it is still highly enjoyable to make your own, choosing your own favorite mixtures of aromatic oils.

Herbs are also good for your hair. Made into rinses and conditioners they can make hair glossy, give it highlights, strengthen the roots and prevent dandruff – besides giving a clean scent! If you wash your hair frequently and need something milder than a commercial shampoo, try using a decoction of soapwort root.

As a finishing touch, add a trace of fragrant toilet water behind your ears to give a hint of natural fragrance all day long.

SCENTED WATERS FOR WASHING

A ewer full of scented washing water was once the ultimate luxury for a lady's bedroom. Sweet washing waters are easy to make and they can be kept in a bottle in the bedroom or bathroom for up to 3 weeks. They can be used cold or mixed with warm water in a basin and they feel best if you use a small natural sponge for washing. Let your skin dry naturally or gently pat it with a soft towel.

Use scented waters without soap as gentle cleansers and refreshers last thing at night or first thing in the morning, rather than for a thorough wash after heavy work. The scent will be lasting but much more gentle and far less cloying than a commercial perfume. Washing waters can also be used in finger bowls at the table.

If you are in a soft water area you can use tap water for making your washing waters, but for the ultimate in softness and freedom from chemicals use a bottled still spring water.

Adding an essential oil to soft water is the easiest way of making a sweet washing water. Waters made in this way also have long keeping qualities.

You will need 6 drops of essential oil to 1 pt (600 ml) of water. Put the water in a bottle larger than its volume. Add the oil and shake well to disperse it. Leave it for 12 hours and shake well before use.

In the sixteenth century, decoctions of herbs and flowers for the bath were mixed with fresh milk as this was thought to have great skin softening properties.

Scents to Use

Which scented oils you choose for washing waters is a matter of taste but it is best if you select the natural and not the synthetic ones. Make up several different waters that you can use according to the time of day and your mood.

Rose alone or rose mixed with lavender is suitable at any time. Lavender alone will ease tension or an aching head and help you to relax; mixed with peppermint or rosemary, it acts as an early morning refresher. You can also use rosemary alone or mix it with marjoram or basil for a clean, fresh, sporty scent. Other plants, such as violet or orange flower, can be used purely for their delicate perfumes.

A decoction of borage leaves, bran and barley or bean meal was recommended in the seventeenth century for adding to the bath water to cleanse and soften the skin.

Washing waters made with fresh or dried herbs

A strong decoction of herbs and flowers will also make an excellent washing water. To 1 pt (600 ml) of spring water you will need 6 tbls chopped fresh herbs or 3 tbls dried; or 1 oz (25 g) fresh flower petals or ½ oz (15 g) dried. If you have enough herbs, make about 3 pt (1.7 liters) at a time. Put the water and herbs or petals into a pan and bring them gently to the boil. Cover and simmer for 30 minutes. Leave them to cool completely and then strain and bottle them. They will keep for up to 2 weeks.

Herbs and flowers to use include rose, lavender, hyssop, lovage (leaves or root), rosemary, marjoram, pennyroyal, lemon balm, costmary and basil. Use them singly or in mixtures. Here are some ideas: lovage alone or rose and lovage (cleansing and deodorizing); rosemary and hyssop (refreshing); lemon balm and lime flower (relaxing); lavender (relaxing and refreshing); lavender and peppermint (refreshing); borage, sage and tansy (cleansing); rose and lavender (sweetly fragrant); rose, lavender and thyme (refreshing and fragrant); costmary and pennyroyal, singly or mixed (fresh, clean scents); chamomile and lemon balm (soothing).

SCENTED WATERS FOR THE BATH

Herbal and floral waters can also be added to the bath water. Using up one brew for each bath is not very practical, so it is best to make them strong and lasting.

For 1 pt (600 ml) water (use spring water if possible, even though you are going to add it to tap water afterwards) you will need 1 lb (450 g) fresh herbs or 8 oz (225 g) dried. The fresh herbs can be in the form of sprigs and the dried ones either sprigs or crumbled leaves, depending on whether you have dried them yourself or bought them.

Put the herbs and water into a pan and bring them gently to the boil. Cover and simmer for 10 minutes. Cool completely. Strain, pressing down on the herbs to extract as much liquid as possible. Add 3 fl oz (85 ml) vodka before bottling. For use, add ¼ pt (150 ml) to each bath.

It is best to use a mixture of herbs, according to preference and availability. Choose from: rosemary, lavender, lemon balm, basil, hyssop, southernwood, chamomile, meadowsweet, elderflower, bergamot, lovage, peppermint, pennyroyal and dried blackberry leaves.

SCENTED BATH OILS

Adding scented oils to your bath water makes your skin fragrant and silky textured, besides filling the whole bathroom with wonderfully scented steam.

Scented bath oils are easy to make. As a base, you will need a pure, odorless oil that will disperse easily in water. The best base oil to use for bath mixtures is called turkey red oil, which is a treated castor oil. This can be bought from pharmacists and some herb specialists. If it is unobtainable you can use almond, avocado or sunflower oil; they do not disperse as well so a preparation made from these must be swished about well in the bath water.

To every 2½ fl oz (65 ml) of the base oil, you will need about 2-2½ tbls of essential oils, either used singly or mixed with other complementing scents. Put the base oil into a bottle. Add the essential oil and shake well. Leave for 2 weeks for the final scent to mature. Shake well before use. Add 1 tbls only to a bath of hot water.

Sweet washing waters were always used by the rich but found a new popularity among all classes after the Great Plague in the seventeenth century when it was gradually realized that cleanliness had a part to play in keeping away disease.

In India, where it is produced, castor oil is used to make an odorless, light-colored soap. The inferior quality castor oils are employed as lamp fuel.

Rose-scented water was commonly found in finger bowls on medieval banqueting tables. A container in the form of a jeweled silver peacock was given to Matilda, wife of Geoffrey of Anjou, for use at table.

Citrus Bath Oil

2½ fl oz (65 ml) base oil

— • —

1 tbls lemon oil

— • —

2 tsp orange oil

— • —

1 tsp lime oil

Rose Bath Oil

2½ fl oz (65 ml) base oil

— • —

2 tbls rose oil

Rose and Lavender Bath Oil

2½ fl oz (65 ml) base oil

— • —

1 tbls rose oil

— • —

1 tbls lavender oil

Spicy Bath Oil

2½ fl oz (65 ml) base oil

— • —

1 tsp cypress oil

— • —

1 tsp sandalwood oil

— • —

1½ tbls clove oil

Rosemary Bath Oil

2½ fl oz (65 ml) base oil

— • —

2 tbls rosemary oil

Floral Bath Oil

2½ fl oz (65 ml) base oil

— • —

1 tbls carnation oil

— • —

1 tbls lavender oil

— • —

½ tbls clove oil

Washing every day with rosemary water and leaving it to dry naturally was once thought to keep skin young looking.

BATH VINEGARS, VINAIGRETTES AND SMELLING BOTTLES

Scented vinegars are refreshing and invigorating. Put into a hot bath they will soften the skin, and made into a cooling compress for the brow they can ease headaches and tension. They can also be dabbed behind your ears and on your temples and used as a skin toner after cleansing. Their refreshing quality is due to the acetic acid which dissolves the aromatic substances in herbs and flower petals, making a perfume of a different composition to one made with alcohol.

Scented vinegars are best made with a good-quality cider vinegar that has a light color and a delicate scent. To scent it, use rose petals, lavender, rosemary, lemon balm, lemon verbena, basil, hyssop, peppermint, scented geranium leaves, jasmine or pennyroyal.

To make a scented vinegar with fresh herbs or petals, half fill a clear glass bottle with the chopped herbs (or whole rose petals). Top up the bottle with

For many centuries, aromatic vinegars were used to ward off infection. In Stuart times doctors carried special walking sticks with a silver knob on the end, inside of which was a vinaigrette made of moss or a sponge soaked in an aromatic vinegar. With this under his nose, the doctor would, it was thought, avoid the infections of his patients.

vinegar. Cover it and leave it on a warm, sunny windowsill for 3 weeks. Strain off the vinegar, pressing down hard on the herbs. Add to it an equal volume of spring water. Bottle and leave for a week.

If you are using dried herbs or petals, put 3 tbls herbs or 6 tbls rose petals into a jug. Mix together ½ pt (300 ml) each of light-colored cider vinegar and spring water. Bring them to just below boiling point and pour them over the herbs. Cover tightly with plastic wrap and leave for 12 hours. Strain and bottle. Larger quantities can be made if wished.

For use, add ½ pt (300 ml) to a bath while the taps are running; or, soak a soft face cloth in the vinegar, wring it out and lay it across your forehead while you relax.

Vinaigrettes and smelling bottles make an excellent addition to the bathroom medicine cupboard. Hold them under your nose to relieve a stuffy head or a headache.

Victorian ladies wore tiny vinaigrettes on chains round their necks.

Vinaigrettes were never more popular than in the Regency period when they were carried by fashionable young men and women meeting in crowded pump rooms and ballrooms.

Vinegar for a Smelling Bottle

5 fl oz (150 ml) light-colored cider vinegar

4 sprigs lavender flowers or 2 tbls dried lavender

4 sprigs rosemary or 2 tbls dried rosemary

2 mint sprigs or 1 tbls dried mint

2 marjoram sprigs or 1 tbls dried marjoram

1 tsp camphorated oil

Put the vinegar and herbs into a bottle. Seal it tightly and leave it in a warm place for 3 weeks, shaking it every few days. Strain the vinegar. Add the camphorated oil and mix well.

Push a small piece of natural sponge into a bottle. Pour in the vinegar and seal tightly.

For use, remove the bottletop and hold the bottle under your nose. Breathe deeply to freshen and clear your head.

Smelling Salts

Fill a small bottle with coarse sea salt. Add 1 tsp tincture of benzoin and 1 tsp lavender oil. Use as the vinaigrette.

BATH-BAGS

When dried herbs and flowers are held under running hot water their scents and properties will be released. Simply putting chopped herbs into a bath, however, will clog up the drain and make the bath extremely difficult to clean, so the answer is to put your herbs into small bags that can be hung underneath a running hot tap. That way, you will get all the advantages and none of the mess! Bath-bags also keep well, enabling you to make a great many at a time so that they will always be on hand.

Bath-bags can be made from muslin, calico or a thin, plain cotton material. For each bag, cut a piece of material 3½ x 9½ inches (9 x 24 cm), using pinking shears if possible as the top edge will not be neatened. Fold the material in half along the short side and machine down each side with a ⅜ inch (1 cm) seam. Clip the corners, turn and press. Run a gathering stitch round the top of the bag, 1½ inches (4 cm) from the edge. Pack the bag with your chosen filling. Pull up the gathers and secure them. Tie the bags round with a long piece of string, cord or ribbon in two knots. Then make a loop by tying the two ends together.

For use, hang the bag on the hot tap of the bath, making sure that it is in the water flow. Alternatively, simply drop a bag into the bath underneath the jet of water from the tap.

Fillings for Bath Bags

You can choose herbs simply for their scent or for their properties, and you can use single herbs or a mixture. Rosemary, meadowsweet, bay, hyssop and mugwort will relieve tired and aching limbs; lemon balm, pennyroyal, lime, chamomile and valerian are soothing; lavender is both relaxing and reviving; borage, sage, tansy and blackberry leaves are invigorating; and lovage is a natural deodorant.

Adding coarse oatmeal or cornmeal to the mixture will soften both the water and your skin.

Here are some suggestions for fillings. Each one will fill 8-10 bath bags made as above. Simply mix the ingredients together before putting into the bags. All herbs are dried.

The Romans used rosemary in baths to relieve tired limbs after a long march. After a battle they added bay leaves, while for everyday use they favored lavender.

Valerian is a relaxing bath herb which, when used in the evening, will help you to a restful sleep. It has a slightly bitter scent, so is best combined with something sweet such as rose petals.

Lime flowers, or a decoction of the flowers, in a bath is an old country remedy for hysteria.

Rose and Lemon Verbena Filling

1 oz (25 g) rose petals

½ oz (15 g) lemon verbena leaves

Rosemary and Peppermint Filling

1 oz (25 g) rosemary

½ oz (15 g) peppermint

4 oz (100 g) coarse oatmeal

Rose, Lavender and Thyme Filling

1 oz (25 g) rose petals

1 oz (25 g) lavender flowers

½ oz (15 g) thyme

4 oz (100 g) coarse oatmeal

BATH MITTS AND WASHING BAGS

A bath mitt or washing bag filled with a herb and cornmeal mixture will, when rubbed gently on the skin, cleanse it and remove any dead cells.

For your own use, make a simple washing bag. Cut 2 × 6 inch (15 cm) squares of muslin. Join them with a ⅜ inch (1 cm) seam, leaving a gap of about 3 inches (7.5 cm) on one side. Trim the corners and turn the bag. Fill it and sew up the gap. The bag can be used 3 times in a hot bath.

Bath mitts make attractive and unusual gifts. To make them, cut out a piece of toweling for the back of the mitt, using the pattern on page 148. Cut 2 pieces of muslin, the same size and shape.

Turn under and press ⅝ inch (16 mm), twice, on the bottom of toweling. Machine stitch it. If wished, lay a strip of ribbon across mitt, 1 inch (2.5 cm) from the bottom. Pin it into place but do not sew.

Turn under and press ⅝ inch (16 mm), twice, on the bottom edge of each piece of muslin. Machine stitch it. Lay the pieces of muslin together with these pressed hems facing inwards. Baste round the sides and top and then lay them on the wrong side of the toweling shape. Pin and machine a ⅝ inch (16 mm)

Washing your face with a decoction of lovage was once thought to lighten freckles. It has mildly antiseptic properties and a strong, fresh scent so it helps to cleanse a spotty skin besides acting as a mild deodorant. A lovage bath will be both refreshing and relaxing.

seam round the sides and top of the mitt. Trim the seam, clipping into the curves and corners. Turn the mitt, keeping the 2 pieces of muslin together, and press it.

Fill the space between the 2 pieces of muslin with the herb and cornmeal mixture. Sew the bottoms of the pieces of muslin together either by hand or by machine, leaving the gap between the muslin and the toweling so that you can put your hand into the mitt.

Remove the pins from the ribbon. If wished, sew a small ribbon bow on to the ribbon. Like the bags, each mitt can be used three times.

Fillings for Mitts and Washing Bags

Mix all ingredients together. All herbs and flowers are dried and each mixture will fill 3 mitts or 4 bags.

Chamomile and Lavender

1 oz (25 g) chamomile flowers

1 oz (25 g) lavender flowers

4 oz (100 g) cornmeal

Peppermint and Elderflower

1½ oz (40 g) peppermint

1½ oz (40 g) elderflowers

4 oz (100 g) cornmeal

Rose, Meadowsweet and Lovage

1½ oz (40 g) rose petals

1 oz (25 g) meadowsweet

½ oz (15 g) lovage root

4 oz (100 g) cornmeal

Whether taken as a tea or put in the bath, chamomile is one of the most relaxing of herbs. The scent of it was once thought to be comforting to the brain.

A bath scented with elderflowers will help to ease irritated or sunburnt skin.

SOAP

Here are two ways of making your own scented soap, using an unscented variety of soap as a base. For the simplest method, scatter 6 drops of essential oil (a single scent or a mixture) over soft cotton fabric like flannelette. Wrap this around a bar of unscented soap. Put into a plastic bag and leave for 2 months. The scent will last as long as the soap and the fabric can be used again.

Molded Rose Soap

This is a simple way of making soap, similar to that used for Elizabethan wash balls.

10 oz (275 g) unscented soap bars

— • —

½ pt (300 ml) rose water

— • —

6 drops pink food coloring

— • —

10 drops rose oil

Grate the soap very finely (there must be no lumps in it as they will not dissolve) and put it into a large mixing bowl. Heat the rose water to just below boiling point and pour it over the soap. Add the food coloring and beat well with a wooden spoon until the mixture is thick and all the soap has melted to make a smooth paste. Add the rose oil and leave the paste for 10 minutes to thicken.

Knead the paste with your hands. Form it into a ball and put it on a smooth, dampened work surface with a sheet of plastic wrap over the top. Roll it to a thickness of about ⅝ inch (16 mm). Cut the paste into small shapes such as hearts, moons, circles etc., with a pastry cutter and, like pastry, form the remainder into another ball and reroll it.

Put the shapes on to a tray or baking sheet which is covered with plastic wrap and leave them for 24 hours. Turn them and leave them for a further 24 hours. They should now be firm.

For your own use, store the soap in a plastic bag. As a gift, arrange the shapes amongst tissue paper in an attractive box.

Variations: Orange flower soap using orange flower water, orange flower oil and orange food coloring; lavender soap using spring water, 12 drops lavender oil and lilac food coloring.

In the eighteenth century, many ladies made their own washballs. *Hannah Glasse*, a well known author writing in 1784, mentions a 'wash-ball scraper', an implement used to polish the dry balls to make them smooth and shiny.

In Georgian times there was a high tax on soaps but this was lifted later for the Victorian gentry who insisted on taking a daily bath.

When soap was scarce and expensive, a variety of surprising ingredients was used to make washballs, including ground raisins and brown breadcrumbs.

Bubble Bath

Put 3 tbls of this under the fast running water from
the bath taps to make a luxurious, foamy bath.

½ pt (300 ml) spring water

4 tbls dried flowers or herbs suitable for
scenting a bath (see above)

½ pt (300 ml) liquid organic cleaner

2 tsps essential oil corresponding with
the dried herbs used

Put the water and herbs or flowers into a pan. Bring
them to the boil. Cover and simmer until the water
has reduced by half (about 30 minutes). Strain,
pressing down well on the herbs or flowers.

Cool the decoction for 10 minutes. Mix it with the
liquid organic cleaner. Add the essential oil. Stir well
and bottle.

SHAMPOOS, CONDITIONERS AND RINSES

Soapwort (*Saponaria officinalis*) has been employed for making soap and shampoo since Anglo-Saxon times when the ground roots were mixed with willow ash. It was frequently used in medieval monasteries. The leaves or roots of the herb will make a foamy liquid when hot water is added, that is similiar to a mild shapoo. Other herbs can be added to the infusion both for their scent and their advantageous properties.

Lemon balm will add a delicious fragance; chamomile, when used frequently, will add highlights to light-colored hair while adding dried marigold flowers to chamomile will give golden highlights; sage is an excellent conditioning herb for darker hair and it also gives a clean scent; rosemary has a strong, sweet aroma and is a good universal conditioner; nettles and elderflowers will combat dandruff.

Rosemary Conditioner for Dry Hair

Rosemary has been known as a hair tonic for many centuries. Besides conditioning the hair it keeps dandruff away and has been used to heal head sores.

Put 1 oz (25 g) dried rosemary leaves into a bottle. Pour in 1 pint (600 ml) olive oil. Seal the bottle tightly and leave it on a sunny windowsill for 3 weeks, shaking gently every few days. Strain the oil and bottle it again.

After washing your hair, rub in 4-6 tbls of the oil. Wrap your hair in a hot towel and leave it for 30 minutes. Rinse out the oil with hot water.

Herbal Rinses

Herbal infusions can be used to rinse hair after washing. They will condition the hair, leaving it shiny and fragrant.

Used 1 oz (25 g) dried herbs to 2 pts (1.15 liters) boiling water. Leave the infusion, covered, until it is cool and then strain it.

After washing you hair, rinse out all the shapoo with warm water and use the infusion for the final rinse, pouring it through your hair several times.

Herbs to use are camomile, alone or mixed with yarrow; sage, alone or mixed with horsetail; rosemary, alone or mixed with nettles, limeflowers or southernwood.

Southernwood once had a reputation for making hair grow. It was called 'lad's love', as a regular application to the chin was thought to promote the growth of a beard.

Soapwort Herbal Shampoo

The soapwort mixture is a very mild one and can be used every day if wished.

½ oz (15 g) soapwort root, crushed
— • —
½ oz (15 g) herbs as above
— • —
2½ pints (1.45 liters) water

Put the soapwort root and the herbs into a pan with the water. Bring them to the boil and simmer them for 4 minutes, covered. Take the pan from the heat and leave, still covered, until the decoction is quite cool. Strain, pressing down well, and bottle.

To use, wet the hair and rub in 7 fl oz (200 ml) of the shampoo. Rinse.

To Make a Dry Shampoo

This is ideal to use when you haven't the time or the facilities to wash your hair. Pack some in your bag when you go camping! It makes your hair feel clean and glossy.

½ oz (15 g) fuller's earth
— • —
½ oz (15 g) orris root powder
— • —
½ oz (15 g) arrowroot
— • —
10 drops of rosemary oil

Mix all the ingredients well together. Use one-third of the mixture for one application; rub it into your hair and leave it for 10 minutes. Brush it out, preferably with bristle brush.

Colognes and Toilet Waters

Colognes and toilet waters are fragrant and refreshing when splashed on your body after a wash or bath.

The use of scented waters and perfumes made by the distillation of flowers became popular in the sixteenth century when Queen Elizabeth set up a still for the use of her ladies at Hawkstead. Within a hundred years they were being distilled from mixtures of flowers, herbs and spices in many country homes. The process is complicated (and also against the law!), so it is much easier nowadays to buy a 70 percent proof spirit or a scentless vodka and add fresh or dried herbs and flowers, spices and essential oils.

One of the simplest methods of making a toilet water or cologne is to add essential oils to alcohol or a mixture of alcohol and spring water. Experiment first, however, by adding the oils of your choice, drop by drop, to plain water until finally you find the scent that suits you. This will save any expensive waste of a badly scented alcohol.

The name 'cologne' comes from the most popular scent of all, Eau de Cologne, which was introduced to Europe in 1725 by Giovanni Maria Farina, an Italian living in Cologne. Its principal ingredients are the oils of bergamot, lavender, lemon, neroli and citron.

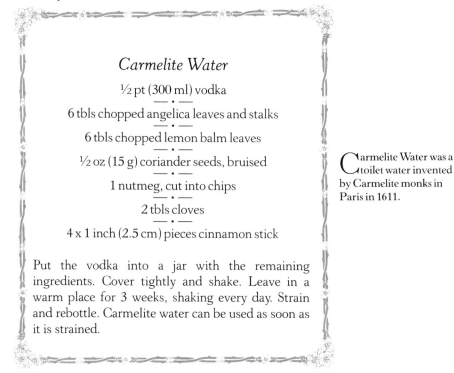

Carmelite Water

½ pt (300 ml) vodka

— • —

6 tbls chopped angelica leaves and stalks

— • —

6 tbls chopped lemon balm leaves

— • —

½ oz (15 g) coriander seeds, bruised

— • —

1 nutmeg, cut into chips

— • —

2 tbls cloves

— • —

4 x 1 inch (2.5 cm) pieces cinnamon stick

Put the vodka into a jar with the remaining ingredients. Cover tightly and shake. Leave in a warm place for 3 weeks, shaking every day. Strain and rebottle. Carmelite water can be used as soon as it is strained.

Carmelite Water was a toilet water invented by Carmelite monks in Paris in 1611.

122

Rose and Lavender Cologne

½ pt (300 ml) 70 percent alcohol or vodka

— • —

3 drops rose oil

— • —

3 drops lavender oil

— • —

2 drops clove oil

Put the alcohol or vodka into a jar or bottle. Add the oils and shake well. Leave for 1 week.

Lavender and cloves complement one another and together they were used in a scent named Rondoletia, after its inventor, a botanical writer of the sixteenth century.

Lavender Water

8 fl oz (250 ml) spring water

— • —

1 oz (25 g) dried lavender flowers

— • —

5 fl oz (150 ml) 70 percent proof alcohol or vodka

— • —

2 drops lavender oil

Put the spring water and lavender into a saucepan. Bring them gently to just below boiling point, cover and simmer for 5 minutes. Take the pan from the heat and cool the contents completely. Strain off the water, pressing down on the lavender.

Put the water into a jar. Add the alcohol and lavender oil. Shake well. Leave for 1 week.

Lavender water was first commercially distilled in the early seventeenth century and the earliest English recipe dates back to 1615.

Spring Flowers Cologne

½ pt (300 ml) 70 percent proof alcohol or vodka

— • —

2 oz (50 g) fresh, sweet-scented rose petals or
1 oz (25 g) dried

— • —

1 oz (25 g) cassia bark

— • —

4 drops violet oil

— • —

4 drops bergamot oil or orange flower oil

Put the alcohol or vodka into a jar with the rose petals and cassia bark. Cover tightly and leave for 1 week in a warm place. Strain, pressing down well on the rose petals. Put the strained alcohol back into the jar. Add the essential oils and shake well. Leave for 1 week.

SCENTED CANDLES AND BURNING PERFUMES

Since ancient times, mixtures of herbs and spices have been burnt to scent and fumigate rooms. The ancient Egyptians, Greeks and, later, the Romans all used burning both in their houses and, most importantly, in places of worship. In fact the word 'perfume' comes from the Latin 'per' meaning through and 'fumare' meaning to smoke.

Burning aromatic herbs and spices was for many years believed to cleanse both physically and spiritually. The rich, aromatic fumes given off from burning pans and open fires drove away infection from those who were sick and protected those who nursed them. Indeed, many incense and chafing dish mixtures are remarkably head-clearing and of great value to sufferers from head colds. Scented candles produce warm, soft scents, vaporizing and chafing dish mixtures are sweet and clean and incense mixtures highly pervasive.

None are hard to make. Even candles, once you have the right ingredients and equipment and you follow the method step by step, are surprisingly easy to make. It is really pleasing to peel away a strangely shaped rubber mold to reveal an intricate shiny candle underneath.

The vaporizing, chafing dish and incense mixtures are all made by simply mixing the ingredients. Follow the recipes given the first time you make them and then experiment with the ingredients that you find the most pleasing.

The dry vaporizing mixtures can be kept on the radiator at all times. Use the chafing dish and incense mixtures to freshen rooms and drive away unpleasant smells.

SCENTED CANDLES

To make scented candles you will need wax, wicks, colorings, candle scents or fresh or dried herbs, and molds. All these, apart from the herbs, can be bought from craft shops.

Wax

The first wax to be used for candles was beeswax which, in its natural state, is a pale amber color and which can now also be bought in a refined white form which it is possible to color. It burns slowly and steadily and gives off a delicious, sweet, honey scent as it burns. You can buy natural beeswax in block form from many beekeepers and both natural and white beeswax in the form of small, flat discs, in drops like chocolate drops or in bars is available from craft shops. Beeswax can be used alone or mixed with paraffin wax.

Most candles today are made with a large proportion of paraffin wax. It is white in color, and is usually bought by weight in the form of small, chocolate-drop shapes or rectangular blocks. It is odorless and sets to give a glossy, translucent finish. Its one drawback is that, when used alone, it will drip too much as the candle is burning. In order to correct this, it is usually mixed with another form of wax called stearin which is made from natural animal and vegetable oils and which comes in the form of small, white granules. Generally, you will need 90 percent paraffin wax and 10 percent stearin.

Do not use stearin with rubber molds as it may damage them. Use wax whitener in the same proportions. If this is not available substitute Vibar (a special wax sold by candle maker's shops) in the proportions of ½ percent to 99½ percent paraffin wax.

Craft shops sell dye discs and dye powders for coloring candle wax. You can also use ordinary wax crayons. The dyes tend to be very strong and only one-sixteenth of a dye disc is needed to color 1 lb (450 g) wax. The dye disc or crayon is mixed with the stearin. Both are melted together over a low heat and then stirred into the melted wax. To test for the final color, drop 1 tsp of the mixed waxes and dye into a saucepan of cold water. It will set to give a color slightly lighter than the finished candle.

Breathing in a mixture of pure, clean scents was thought to cleanse the mind, hence the use of incense in places of worship both in ancient temples and in Christian churches. The heady scents were also slightly intoxicating, inducing high religious excitement.

Scenting the Candle

Candles can be scented either with herbs or with special oils.

Dried herbs can actually be stirred into the melted candle wax so they become part of the composition of the candle and give a strong, fresh scent as it burns. If the candle wax is left uncolored, it will take on a translucent green shade. You can also tint this type of candle a similar color to that of the herb flowers. Some herbs will sink to the bottom of the candle, others will float and some will distribute themselves more evenly. No two will be the same.

Fresh sprigs of strongly scented herbs, such as rosemary, lavender, lemon verbena or hyssop, will give the candle a mild fragrance if they are infused in the melted wax at a temperature of 180°F (82°C) for 45 minutes.

Craft shops specializing in candle making ingredients supply a wide range of special oils for candles. Do not use ordinary oils left over from pot-pourri making as they may not distribute evenly through the wax.

When you are scenting candles with a high proportion of beeswax, take account of the honey scent of the wax and choose oils that will complement it; all herbs blend beautifully with it.

In eighteenth-century America, wax extracted from plants was used instead of animal fats. The plant employed depended upon its availability in a certain area. Waxberry was used in the South, candleberry in the East and bayberry in the American colonies.

Candle Molds

Candle molds are available in clear plastic, glass, metal or rubber. Candles made by infusing bunches of fresh herbs in the wax or by adding chopped herbs can safely be made in all types of mold. A wax scented with a special candle perfume must be put in metal or rubber molds only, since it may cause the plastic molds to bubble or crack.

The scent of bayberry is clean and spicy and bayberry perfume is still popular, particularly for Christmas candles.

Plastic, glass and metal molds are made in smooth sided, plain shapes (cylindrical, square, conical etc.) to enable the finished candles to slip out easily. The flexible rubber molds can be peeled off and so tend to be much more intricate, either with relief designs on cylindrical candles or in a variety of novelty designs such as flowers, vegetables or small figures.

127

The plastic molds are made so that they can stand upside down without support for taking the melted wax. Metal molds come with their own stands, but separate stands must be bought for glass and rubber molds.

You can also improvise by using yoghurt or cream containers, tin cans or cardboard tubes, but candles made with these have flat tops and do not burn as well as the pointed ones that are produced by special molds. Before making your candles, measure the capacity of your molds and use only as much wax as will fill them. This will avoid waste.

Generally, 12 oz (350 g) paraffin or beeswax (or a mixture) plus 1¼ oz (35 g) stearin will make a total volume of ¾ pt (450 ml).

Wicks

Candle wicks are made from bleached cotton yarn, woven into varying thicknesses. In most cases, the wick is named after the diameter of the candle that it best suits, so a 1½ inch (4 cm) wick, for example, should be used for a candle of a diameter of 1½ inches (4 cm). When using a high proportion of paraffin wax, you may need a slightly thinner wick. Beeswax candles need slightly thicker wicks.

If you are using rubber molds you will need a wicking needle to thread the wick through.

To seal the hole in the mold through which the wick protrudes, you will need a special mold seal or plasticine.

Basic Method For Making Candles

Prepare the molds by lightly oiling them with a good quality cooking oil such as sunflower oil.

Push the wick through the hole in the top of the mold and bring it through to the rim. Tie it on to a cocktail stick or a knitting needle which you then lay across the rim of the mold, keeping the knot in the center. Pull the wick taut at the other end and anchor it with mold seal, plasticine or Blu-tack to cover the hole completely and prevent the wax from dripping through. Trim off the spare wick, leaving about 1 inch (2.5 cm) protruding through the hole. Place the mold on its stand.

Put the wax into a double boiler or in a bowl in a pan of water, melt it on a low heat and bring the temperature to 180°F/82°C. If you are scenting the candle by infusion add the sprigs of fresh herbs now, and maintain the temperature for 45 minutes. Remove the herbs before proceeding. Add dried herbs at this point.

Melt the stearin in another small double boiler or pan. Melt the color with it, if you are tinting the candle. Stir into the wax.

Pour the wax mixture into the molds. Keep any remaining wax warm but do not boil. Leave the wax for 10-20 minutes, depending on the size of the candle. As the wax cools it will contract and a depression will form in the center. Carefully break the surface with a cocktail stick and top up with more wax (you may have to do this more than once). After

The second day of February is Candlemas, when it was an old English custom to take candles to church to be blessed.

The color of religious candles was once significant. White meant purity and innocence and white candles were also burned at a time of religious reaffirmation such as at a baptism, at communion and on Christmas and Easter day.

the final topping up, leave the candles for about 5 hours or until they are set completely.

To remove straight candles from the molds, remove the mold seal and then gently pull on the knitting needle or cocktail stick. Rubber molds should be gently peeled off.

Trim the wicks to about ½ inch (1.5 cm).

Polish the outside of the candles with cotton balls or wool which you have dipped in vegetable oil.

To clean your pans and basins, tip away any unused wax (into a bin, not down the drain!), pour in boiling water and scrub with a long-handled brush.

Make all the following candles according to the basic method.

Candles made with stearin burn well and remain steady and upright. Pure paraffin wax candles may bend in the heat from their own flame.

Pure Beeswax and Herb Candles

12 oz (350 g) white beeswax

— • —

2 sprigs rosemary

— • —

4 sprigs hyssop

— • —

6 lavender spikes

Following the basic method, infuse the herbs in the melted wax and removing them before pouring the wax into the mold

Herb candles should be a very pale color to allow the color of the herb leaves to shine through the wax.

Beeswax and Lemon Verbena Candles

8 oz (225 g) white beeswax

— • —

4 oz (100 g) paraffin wax

— • —

1¼ oz (35 g) stearin

— • —

tiny amount of grated yellow dye disc
to make a very pale colour

— • —

1 oz (25 g) lemon verbena leaves, crushed and with central stems removed

131

Floral Candles

12 oz (350 g) paraffin wax

—·—

1¼ oz (35 g) stearin or wax whitener

—·—

⅛ tsp pink dye disc, grated

—·—

⅛ tsp orange dye disc, grated

—·—

4 drops floral candle perfume, or 2 drops
honeysuckle and 2 drops violet perfume

R ed candles stand for joy, love and life; and green symbolize new life, hope and understanding.

Spiced Candles

12 oz (350 g) paraffin wax

—·—

1 ¼ oz (35 g) stearin or wax whitener

—·—

1/16 brown dye disc, grated

—·—

4 drops sandalwood perfume

—·—

2 drops patchouli perfume

In the seventeenth century, scented candles were made by mixing a variety of spices plus dried and ground rose petals and lemon peel held together with gum tragacanth.

Apple Candles

6 oz (175 g) white beeswax

—·—

6 oz (175 g) paraffin wax

—·—

1¼ oz (35 g) wax whitener

—·—

⅛ apple candle disc (contains color and perfume)

Tallow candles are still made in rural Spain. Animal fats are mixed with lime which dissolves away any impurities and the fat is then simmered for several hours with vinegar before being molded or used for dipping.

Beeswax and Rosemary Candles

8 oz (225 g) white beeswax

—·—

4 oz (100 g) paraffin wax

—·—

1¼ oz (35 g) stearin

—·—

⅛ green dye disc, finely grated

—·—

1 oz (25 g) dried rosemary leaves

Decorating Candles

Candles can be decorated with pressed sprigs of herbs and flowers, with poster paints and with transfers.

For pressed flowers, keep back a little of the wax when you are making your candle. After the candle has come out of the mold, re-melt the remaining wax and paint a little on the back of the flower or herb. Press this on to the candle, paint it over with a little more wax to keep it in place and glaze it.

Ordinary poster paints can be used to paint candles if the paint is first mixed with a little liquid soap, about 1 tsp soap to every 1 tbls paint.

Special candle transfers in various designs can be bought from craft shops.

VAPORIZING

When a gentle heat is applied to essential oils, and also to strongly scented herbs and spices, an aromatic scent is released into the air. You can buy vaporizers for essential oils, or you can make your own.

Vaporizing with a Candle Flame

You will need a nightlight or a 1 inch (2.5 cm) stub of a thick candle. Put this into a pottery nightlight holder or a round pottery jar that has a pattern of holes cut into its sides. Fill a small, heatproof dish with water and 2 or 3 drops of essential oil. Very often a mixture of oils is best, such as 2 drops of rose to 1 of lavender or 2 drops of lemon verbena and 1 of rosemary. Light the nightlight or candle and place the dish of water and oil over the opening in the top of the holder or jar. The vapor from the oils will be released into the room as soon as the flame is lit.

Check about every 30 minutes to make sure that the water is not drying up.

Mixtures for Placing over Radiators

Strongly scented mixtures of herbs, spices and essential oils can be placed in small amounts in dishes on metal shelves above radiators. They can also be put into small sachets and hung over the radiator itself. Use the lavender bag pattern given on page 87, choosing fabrics that complement your curtains and furniture, and tie two bags together with ribbon or cord so that one hangs on either side of the radiator.

When in use these mixtures will last for several months and their scent can then be revitalized by adding more essential oils. When the mixtures are stored in a plastic bag, however, their scent will last for up to a year, so you can make fairly large amounts of the mixtures and store them.

B ecause so many candles were used by the church in the sixteenth century, they were scarce and often too expensive for ordinary people to buy. Consequently there were unwritten rules as to when they were permitted to be used. Shepherds were allowed candles for their lanterns at lambing times and monks could carry them between their dormitory and the chapel when attending services during the night.

Sweet Lavender Vaporizing Mixture

2 oz (50 g) lavender

1 oz (25 g) rose petals

1 oz (25 g) cloves, bruised

1 oz (25 g) cinnamon sticks, crushed

1 oz (25 g) orris root powder

1 oz (25 g) gum benzoin powder

2 tsp rose oil

2 tsp lavender oil

2 tsp clove oil

1 tsp tincture of benzoin (optional)

Put all the dry ingredients into a bowl. Add the oils and tincture of benzoin and mix with your hands.

The scent will be very 'raw' at first, so put the mixture into a plastic bag, seal it and leave it for 3 weeks before using.

Rosemary and Lemon Balm Vaporizing Mixture

2 oz (50 g) rosemary

2 oz (50 g) lemon balm, crumbled

½ oz (15 g) bay leaves, crumbled

½ oz (15 g) *Eucalyptus citriodora* leaves, crumbled (if available)

1 oz (25 g) cloves, bruised

1 oz (25 g) orris root powder

1 oz (25 g) musk powder

2 tsp rosemary oil

2 tsp lemon balm oil

1 tsp clove oil

Make as for Sweet Lavender mixture.

BURNING IN INCENSE HOLDERS

Simple mixtures of herbs such as rosemary or angelica seeds and sugar were once kept burning in perfume pans over open fires in wayside inns to keep the rooms fresh and ready for guests at all times. Such perfume pans are no longer available, but you can buy brass incense burners which are usually imported from the Middle East and India.

Putting charcoal in the burner with the herb or spice mixture will enable it to continue smoldering for a long time. Use discs of instant burning charcoal that are available from herb specialists or the charcoal nuts that are sold for barbecues. In a small incense burner about 2½-3 inches (6-7.5 cm) in diameter you will need about one quarter of a disc 1 inch (2.5 cm) in diameter or a piece of barbecue charcoal of equivalent size.

Before any burning mixtures are put into the burner the charcoal must be gently smoldering and covered with a layer of gray ash. To achieve this, either put it into the burner with a small blob of

The coronation of King George III was attended by the King's Groom of the Vestry, dressed in scarlet livery and carrying a perfume pan of burning herbs and spices.

An old French name for rosemary was *incensier*, since cheap rosemary from the herb garden was often used in churches instead of costly incense.

Gum Benzoin Mixture 1668

1 oz (25 g) gum benzoin powder

2 tsp musk powder

2 tsp granulated sugar

Juniper Incense 1662

1 oz (25 g) juniper berries, well crushed

2 tsp gum benzoin powder

2 tsp musk

3 drops lemon oil

3 drops clove oil

3 drops rose oil

Rose Incense

1 oz (25 g) rose petals, reduced to a coarse powder in a food processor or liquidizer

1 oz (25 g) frankincense powder

1 oz (25 g) myrrh powder

1 oz (25 g) sandalwood powder

6 drops rose oil

Alternative: for a fresher, herbal scent add 4 tbls rosemary and 3 drops rosemary oil.

barbecue lighting fuel or with a piece of paraffin fire lighter about equal in size to the charcoal. Light the fuel or lighter with a match and by the time it has burned away the charcoal will be smoldering.

Spoon 1-2 tsp of a chosen herb or spice mixture over the charcoal. Put on the lid of the burner and leave the charcoal to smolder gently.

Larger burners need larger pieces of charcoal and a larger amount of mixture.

Since the time of the ancient Egyptians, frankincense has always been the most important ingredient in all incense used for religious purposes.

Herb and Sugar Mixtures

You will need crumbled dried herbs such as rosemary, lavender, lemon balm, lemon verbena, southernwood and thyme or crushed angelica seeds and, to each 3 tbls of these, 1 tbls of either white or brown sugar. You can also add a few drops of a flower or herb oil.

Put a piece of charcoal disc into the incense burner and light it. When it is gently smoldering, scatter on 2 tsp of the mixture. Blow out any high flames and let the mixture smolder.

Spice Mixtures

These spicy incense mixtures are based on recipes from the seventeenth century. Simply mix all the ingredients together and leave them in a sealed plastic bag for 2 weeks to mature.

BURNING ON AN OPEN FIRE

Dried sprigs of herbs, herb seeds and large leaves such as bay will all perfume a room if they are placed directly on an open fire when the flames are low. A wood fire is best as wood can add its own scents to those of the herbs.

Rosemary has a strong, clean scent and it was once a favorite for burning in sickrooms. Lavender fills the room with an incense-like smell. Angelica seeds have a sweet scent and were often used in damp and musty rooms.

Dried roots of elecampane give a violet scent; bay smells fresh and clean; southernwood has a fragrant, grassy scent.

Aromatic woods such as juniper and cypress were often burned to fumigate rooms in the sixteenth century. They were sold by the pound by pedlars. Juniper was still favored by students at Oxford a hundred years later.

Scented Pine Cones

Pine or fir cones burn well and give off a faint scent of pine. If you drip a little essential oil on to each one they will perfume your room twice. Leave them in a basket near the fire, so that the scent will be released as the room warms up. Then, to intensify the scent, drop a cone on to low flames about every 30 minutes.

The herb, spice and wood oils, such as sandalwood, cedarwood, clove, cinnamon, lavender, lemon balm and rosemary, are the best to use.

Burning rosemary was said to ward off evil spirits and burning southernwood was believed to drive away 'serpents lurking in corners'.

CHAFING DISHES

In the days when there were open fires in every room, dishes or special perfume pans of spiced and scented water were placed over the fire and, as they simmered, their fragrances were released.

Few open fires these days have stands for pans and pots, but if you simmer the mixtures gently on the kitchen stove and keep all the inside doors open and the outside doors shut, a gentle fragrance should pervade the whole house.

To use any of the following mixtures, simply put them into a small frying pan and bring them gently to the boil. Let them simmer until the water has almost all evaporated, but do not let them boil away completely as this makes a sticky, burnt mess in the pan. The scent will be very gentle at first, but it gradually increases as more water is evaporated.

Wormwood was once burned to keep away fleas. Rue was thought to ward off infection at times of illness.

Queen Elizabeth I's Chafing Dish Mixture

4 fl oz (120 ml) plain water

— • —

2 tsp granulated sugar

— • —

4 tbls dried marjoram

— • —

1 tbls gum benzoin powder

King Edward VI's Chafing Dish Mixture

3 fl oz (85 ml) rose water

— • —

2 tsp granulated sugar

Chafing Dish Mixture From 1662

4 fl oz (120 ml) rose water

— • —

2 tsp granulated sugar

— • —

2 tsp ground cloves or 2 tbls juniper berries, crushed

FRAGRANCES FOR CHRISTMAS

What better time than Christmas to fill the house with scented decorations and also to make gifts of the fragrant items that you have produced throughout the year?

All the scented things described in this book make ideal presents and, because you have made them yourself, they will be much more special than bought presents, however expensive. You can also make them for your own use at Christmas, to give your house a warm, welcoming, spicy fragrance.

Any deep red petals, such as those of poppy or peony, can be made into a pot-pourri with bay leaves and warm scented spices.

Use spices again to make braided spice ropes in the seasonal colors of red and green, or to fill decorations made of cotton material with a Christmas motif.

Hang scented fir or pine cones on the Christmas tree or burn them on an open fire and sit in the glow of a gently fragrant candle.

Red flowers that have been dried whole can be made into posies and tied with Christmas ribbon. Use them alone, with sprigs of evergreen or with bought dried flowers that have been dyed green, and scent each flower with a drop of fragrant oil.

Whole flowers, again scented with oils, can be stuck into a round ball of dried flower oasis. Use a ball about 3 inches (7.3 cm) in diameter and, to hang it, a 30 inch (76 cm) length of ¾ inch (2 cm) wide red or green ribbon. Fold the ribbon in half and push the two cut ends through the ball with a skewer. Pull them through, leaving a loop at the top for hanging, tie the bottom ends together and cut them into V-shapes. Use predominantly red dried flowers, interspersed with green, white and yellow, cut them with a stalk of about 1 inch (2.5 cm) and push them into the oasis.

A Christmas Pot-Pourri

A pot-pourri with rich red colors and a warm spicy scent of the holiday season.

1 oz (25 g) bay leaves, crumbled
— • —
1 oz (25 g) poppy petals
— • —
1 oz (25 g) hibiscus flowers
— • —
1 oz (25 g) peony petals
— • —
2 oz (25 g) cloves, bruised
— • —
1 oz (25 g) cinnamon sticks, broken
— • —
½ oz (15 g) sanderswood
— • —
1 oz (25 g) orris root powder
— • —
6 drops clove oil
— • —
3 drops orange oil
— • —
3 drops lemon oil

Mix all the dry ingredients. Add the oils and mix well. Store in a sealed plastic bag for 6 weeks before using the pot-pourri.

Branches of bay were used as Christmas decorations in England from medieval times to the seventeenth century. They were thought to protect houses from witchcraft and evil spirits.

Christmas Spice Rope

Hang anywhere around the house for decoration and for a spicy scent. Makes one spice rope.

9 × 50 inch (127 cm) lengths of red or green tapestry wool, plus extra for tying
— • —
1 × 2½-3 inch (6-7.5 cm) diameter wooden curtain ring
— • —
4 × 6 inch (15 cm) diameter circles of cotton material printed with a Christmas design
— • —
6 tbls Christmas spice mixture (see following pages)
— • —
3 red or green dried flowers, or sprigs of holly or mistletoe

Make according to the directions for spice ropes given on page 97.

Christmas Candles

12 oz (350 g) paraffin wax

— • —

1¼ oz (35 g) stearin or wax whitener

— • —

¹⁄₁₆ red or Christmas dye disc

— • —

1 tsp bayberry candle perfume

— • —

½ tsp sandalwood candle perfume

Make the candles, following the directions on page 130. You can use plain molds and decorate them later with transfers or by painting; or you can use rubber molds made in special Christmas patterns. Those in the picture depict the *Twelve Days of Christmas*.

Fir branches and small fir trees were brought into the house for Christmas in Germany in the eighth and ninth centuries. Prince Albert introduced Christmas trees to England in the nineteenth century.

The Twelve Days of Christmas

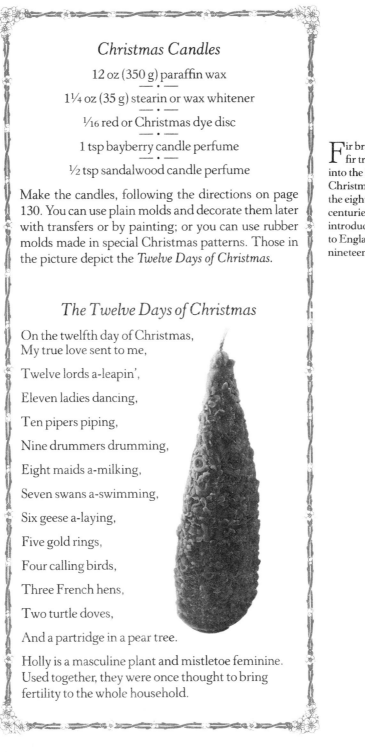

On the twelfth day of Christmas,
My true love sent to me,

Twelve lords a-leapin',

Eleven ladies dancing,

Ten pipers piping,

Nine drummers drumming,

Eight maids a-milking,

Seven swans a-swimming,

Six geese a-laying,

Five gold rings,

Four calling birds,

Three French hens,

Two turtle doves,

And a partridge in a pear tree.

Holly is a masculine plant and mistletoe feminine. Used together, they were once thought to bring fertility to the whole household.

Scented Pine Cone Tree Decorations

These pine cones are simple and cheap to make. They look exceptionally pretty on a tree that is decorated with red and green and will give a delicious scent to a warm room.

fir or pine cones

— • —

wire

— • —

white poster paint

— • —

essential oils eg: clove, sandalwood, lemon, orange

— • —

red and green ¼ inch (6 mm) ribbons

Hold the fir cone stalk end up. Wind the wire round and under the 'petals' about 2 layers down from the stalk. Bring it over the base of the stalk to make a handle. Loop the end through the first ring and twist it.

Using white poster paint, paint all over the stalk end of the cone, including the wire. Dab paint on the rest of the cone so it looks frosted but not completely covered. Do the same with the other cones and leave them to dry.

Dip a clean fine paint brush in essential oil and paint under the petals of the cones. Put the cones into a plastic bag and leave them for 2 weeks. Thread red or green ribbon through the handles and tie it to make a loop to hang on the Christmas tree.

Spice-Filled Tree Decorations

For these you will need a small length of Christmas patterned cotton fabric plus, for each decoration, 5 inches (13 cm) red or green ¼ inch (6 mm) ribbon.

Using templates (see page 147), cut out as many shapes as you will need. Place them together in pairs, right sides together.

Fold each piece of ribbon in half. Place the ribbon loops between each pair of shapes, cut ends facing outwards, in the top center of the bells and in the top back corner of the stocking.

Machine stitch all round the pairs of shapes, leaving a small gap for turning and filling and taking care not to catch in the ribbon loop. Clip into the curves and snip the corners.

Turn and press the shapes. Fill them with the spice mixture and sew up the gaps by hand.

Christmas Spice Mixture

(Fills 6 bells, 6 stockings and 4 spice ropes)

½ oz (15 g) lemon verbena, crumbled

— • —

2 oz (50 g) cloves, crushed

— • —

1 oz (25 g) bayberry powder

1 oz (25 g) sanderswood

— • —

1 oz (25 g) orris root powder

— • —

6 drops clove oil

— • —

4 drops lemon verbena oil

— • —

4 drops sandalwood oil

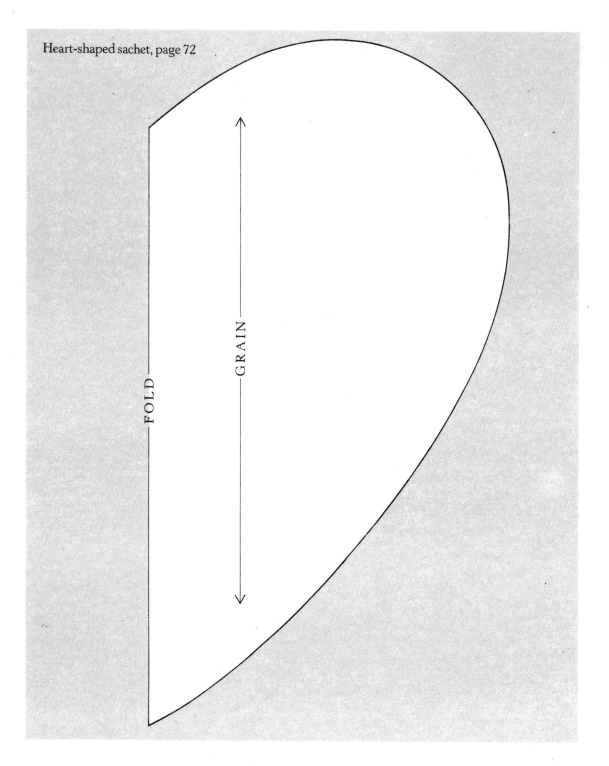

Heart-shaped sachet, page 72

FOLD

GRAIN

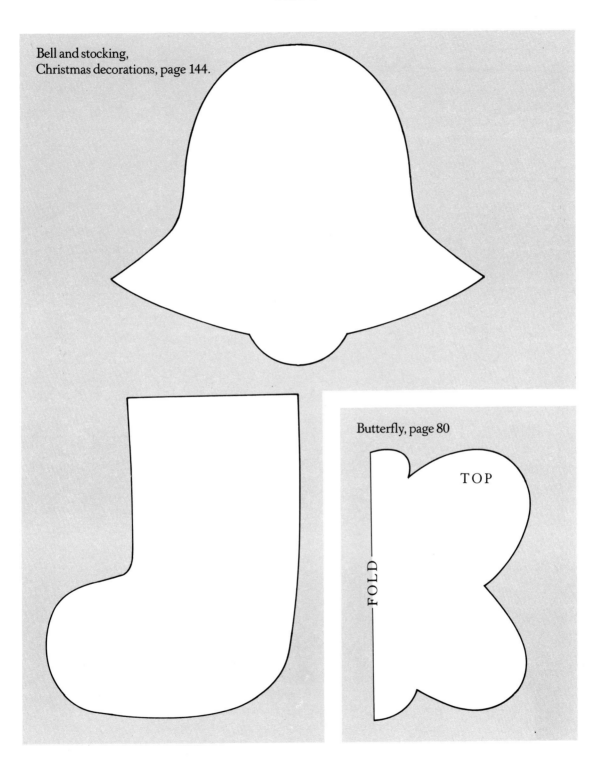

Bell and stocking,
Christmas decorations, page 144.

Butterfly, page 80

TOP

FOLD

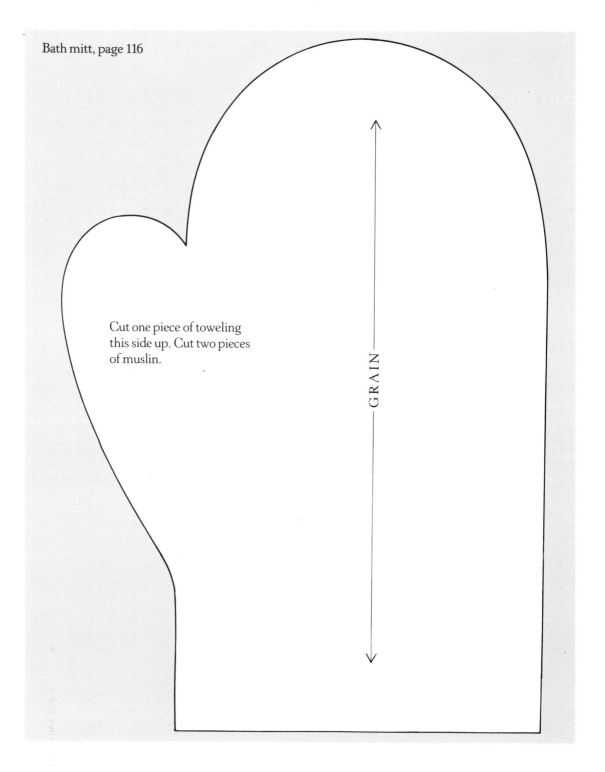

Bath mitt, page 116

Cut one piece of toweling
this side up. Cut two pieces
of muslin.

GRAIN

STOCKISTS AND SUPPLIERS

United Kingdom

All these places will supply mail order. Send sellf-addressed stamped envelop for price lists.

Spices, dried herbs and flowers, woods, barks, roots, essential oils:

G. Baldwin and Co.,
173, Walworth Rd, London,
SE17 1RW

Candles:

Candlemakers' Supplies,
28, Blythe Rd, London, W14 0HA

Herb plants and seeds:

Iden Croft Herbs,
Frittenden Rd, Staplehurst, Kent,
TN12 0DH

Hollington Nurseries Ltd,
Woolton Hill, Newbury, Berks.
RG15 9XT

Roses:

Just Roses,
Beales Lane, Northiam, Rye, Sussex,
TN31 6QY

For unscented washing powders and liquid organic cleaner:

Amway (UK) Ltd,
Snowdon Drive, Winterhill,
Milton Keynes, Bucks. MK6 1AR

U.S.A. and Canada

Dried botanicals

Aphrodisia
282 Bleecker Street, New York,
NY 10014

Caswell-Massey Co. Ltd.
Catalogue Division, 111 Eighth Avenue, New York,
NY 10011

Cherchez
862 Lexington Avenue, New York,
NY 10021

Country Herbs
Box 1133 Stockbridge, MA 01262

Home Fragrance Stores

Agraria
1156 Taylor, San Francisco, CA 94108

Crabtree & Evelyn
30 E. 67th Street, New York, NY 10021

Mail-order sources for plants, seedlings, and seeds

Alberta Nurseries & Seeds Ltd.
P.O. Box 20, Bowden, Alberta, Canada
T0M OK0

Anastasia's Fancy
7231 Central Avenue, St. Petersburg,
FL 33710

Apple Butter Herb Farm
Old Westmoreland Road, Spofford,
NH 03462

Azure & Emeralds Aromatics
P.O. Box 93125, Milwaukee, WI 53203

Back to Eden Gardens
322 East Pole Road, Lynden,
WA 98264

W. Atlee Burpee Co.
300 Park Avenue, Warminster,
PA 18974

Butterflies 'n Blossoms
Route 1, P.O. Box 236, Fayetteville,
TN 37334

Capriland's Herb Farm
Silver Street, North Coventry,
CT 06238

Carroll Gardens
P.O. Box 310, 444 East Main Street,
Westminster, MD 21157

Ceramics by Bob
901 Sheridan Drive, Lancaster,
OH 43130

Clement Herb Farm
Route 6, P.O. Box 390, Rogers,
AZ 72756

Cottage Herb Farm Shop
311 State Street, Albany, NY 12210

Country Manor
Route 211, Box 520, Sperryville,
VA 22740

Ebert Herb & Wheat Farm
Route 1, P.O. Box 281, St George,
KS 66535

Flower Valley Herb Parfumerie
Route 2, P.O. Box 22, Dixon, IA 52745

Frog Park Herbs
RD 2 Frog Park Road, Waterville,
NY 13480

Gatehouse Herbs
98 Van Buren Street, Dolgeville,
NY 13329

Gilbertie's Herb Gardens
Easton, CT 06612

Good Scents
P.O. Box 854, Rialto, CA 92376

Greenfield Herb Garden
Depot & Harrison, Box 437,
Shipshewana, IN 46565

Griffin's
Route 7, P.O. Box 958, Greensboro,
NC 27407

Hartman's Herb Farm
Old Dana Road, Barre, MA 01005

Heartscents
P.O. Box 1674, Hilo, HI 96721

Heirloom Gardens
P.O. Box 138, Guerneville, CA 95446

The Herb Cottage
Washington Cathedral, Mount Saint
Alban, Washington, DC 20016

The Herb Patch
Route 3, P.O. Box 236, St. George, KS
66535

Herb Products Company
11012 Magnolia Blvd., P.O. Box 898,
N. Hollywood, CA 91601

The Herb Connection
2627 John Petree Road, Powder
Springs, GA 30073

The Herbiary & Potpourri Shop
P.O. Box 543, Child's Homestead
Road, Orleans, MA 02332

Herbiforous
Route 1, Elkhart Lake, WI 53020

Herbs in Our Life
118 Cherry Street, Lafayette,
LA 70506

Herbs 'n' Spice
P.O. Box 3358, Sparks, NV 89432

Herbs Unlimited
Harborplace, 301 Light Street,
Baltimore, MD 21202

Hidden Hollow Herbiary
N88 WI8407 Duke Stret, Menomonee
Falls, WI 53051

Hilltop Herb Farm
P.O. Box 1734, Cleveland, TX 77327

Hoo Shoo Too Herb Farm
20261 Hoo Shoo Too Road, Baton
Rouge, LA 70817

Lady Victoria's Flower/Gift Shop
24 Franklin, Clifton, NJ 07011

Lavender Hill Herbs
R 1, P.O. Box 246, Baseline Road,
Kingston, IL 60145

Maine Balsam Fir Products
P.O. Box 123, West Paris, ME 04289

Earl May Seed & Nursery
208 North Elm, Shenandoah, IA 51603

Meadowbrook Herb Garden
Route 138, Wyoming, RI 02898

Meadowsweet Herb Farm
Shrewsbury, VT 05738

Merry Gardens
P.O. Box 595, Camden, ME 04843

Monk's Hill Herbs
RFD 2, Winthrop, ME 04364

Nature's Herb Company
281 Ellis Street, San Francisco,
CA 94102

New York Botanical Gardens
Southern Blvd. at 200th Street, Bronx,
NY 10458

Olde English Gardens Herb House
Shoppes of Wyeth Green in Lightfoot,
Route 60 West, P.O. Box 257,
Williamsburg, VA 23187

Old Sturbridge Village
Sturbridge, MA 01566

Park Seed Co.
P.O. Box 46, Greenwood, SC 29647

Plants of the Southwest/Seeds
1570 Pachebo Street, Santa Fe,
NM 87501

Richters
Goodwood, Ontario L0C 1A0,
Canada

St. John's Herb Garden
5525 Decatur Street, Bladensburg,
MD 20710

The Sassafrass Hutch
11880 Sandy Bottom, NE, Greenville,
MI 48838

Shady Hill Garden
821 Walnut Street, Batavia, IL 60510

Sinking Springs Herb Farm
234 Blair Shore Road, Elkton,
MD 21921

Smile Herb Shop
4908 Berwyn Road, College Park,
MD 20740

Stillridge Herb Farm
10370 Route 99, Woodstock,
MD 21163

Summit Herb Gardens
RD 1, P.O. Box 221, Green, RI 02827

The Sunny Window
P.O. Box 3125, Sax Station,
Framingham, MA 01701

Tansy Farm
RR 1, Agassiz, BC, Canada V0M 1A0

Thompson and Morgan, Inc.
Jackson, NJ 08527

Tom Thumb Workshops
P.O. Box 322, Chincoteague, VA 23336

Treasure Mart Mail-Order
6121 Adkins, St. Louis, MO 63116

Wayside Gardens
Hodges, SC 29695

Well-Sweep Herb Farm
317 Mt. Bethel Road, Port Murray,
NJ 07865

Whipple House Mail Order
59 Fisher Street, Westborough,
MA 01581

White Flower Farm
Litchfield, CT 06759

New Zealand

Hillside Herbs Ltd.
166 Fairy Springs Road, Rotorua
(073) 479 535

Floriste Addingtown
292 Lincoln Road, Addington,
Christchurch
(03) 384 017

Gail's Floral Studios
Centreplace, Victoria Street, Hamilton
(071) 393 758

Australia

New South Wales

The Fragrant Garden
25 Portsmouth Road, Erina NSW 2250
(043) 67 7322

The Lavender Patch
Lot 3, Cullens Road, Kincumber
NSW 2250
(043) 68 1233

H.E. Koch & Co Pty Ltd
1 Probert Street, Camperdown
NSW 2050
(02) 519 8044

Dural's Colonial Cottage & Gallery
62 Kenthurst Road, Dural NSW 2158
(02) 654 1340

Roy H Rumsey Pty Ltd
P.O. Box 1, 1335 Old Northern Road
Dural NSW 2158
(02) 652 1137

Swanes Nursery
490 Galston Road, Dural NSW 2158
(02) 651 1322

Argyle Soap & Candle Co.
33 Playfair Street, The Rocks
NSW 2655
(02) 241 3365

Common Scents Herb Cottage
745 Old Northern Road, Dural
NSW 2158
(02) 651 1027

Aguis Phillip
Suite 67, 61 Marlborough Street,
Surrey Hills NSW 2010
(02) 690 1703

The Flower Warehouse
Cnr Barney and Castle Streets, North
Parramatta NSW 2150
(02) 630 7466

Victoria

Ring of Roses
90 Maling Road, Canterbury Vic 3126
(03) 836 2814

Potpourri & Sachet Supplies
P.O. Box 53, Northcote Vic 3070
(03) 489 8405

Coora Cottage Herbs
Thompsons Lane, Merricks Vic 3916
(059) 89 8338

The Gumleaf Candle Co.
No 80, Mangans Road, Lilydale
Vic 3755
(03) 735 3755

Australian Botanical
68 Burwood Road, Hawthorn Vic 3122
(03) 818 2673

Queensland

Ahisma
Drivers Court, Cobble Creek Qld 4523
(07) 289 9191

ACKNOWLEDGEMENTS

We should like to thank:
Martin and Judith Miller for generously allowing us to use Chilston Park Country Hotel, Lenham, Kent, plus its many contents, for all our photography sessions, and Sue Greenwood for setting it all up for us; Rosemary Titterington of Iden Croft Herbs for all her help, ideas and advice;

Judith and Simon Hopkinson of Hollington Herbs for allowing us to photograph their herbs and for information about *Iris germanica*; Just Roses for pointing out the most fragrant blooms in their catalogue;

Peckwater Antiques, Charing, Kent, for the loan of various props for photography;

Eric and Hazel Wolfson of the Dower House, Chilston Park, for allowing us to pick their roses;

Mary Turner of Faversham, Kent, for the wonderful idea of the knitted sachets and for making the sachets for the picture, including spinning and dyeing the wool;

Vic Morris for making the drying frames; Jenny Hudson, for use of her dining room and bowls.

151

INDEX

Page numbers in *italics*
refer to illustrations.